The New Knowledge Ec  pe

The New Knowledge Economy in Europe

A Strategy for International Competitiveness and Social Cohesion

Bengt-Åke Lundvall, Gøsta Esping-Andersen, Luc Soete, Manuel Castells, Mario Telò, Mark Tomlinson, Robert Boyer, Robert M. Lindley

Edited by

Maria João Rodrigues

Professor, University of Lisbon – ISCTE and Special Advisor to the Prime Minister, Portugal, Chair of the Advisory Group of Social Sciences in the European Framework Programme, EU.

Edward Elgar
Cheltenham, UK • Northampton, MA, USA

Published by
Edward Elgar Publishing Limited
Glensanda House
Montpellier Parade
Cheltenham
Glos GL50 1UA
UK

Edward Elgar Publishing, Inc.
136 West Street
Suite 202
Northampton
Massachusetts 01060
USA

Paperback edition 2003

A catalogue record for this book
is available from the British Library

Library of Congress Cataloguing in Publication Data

The new knowledge economy in Europe: a strategy for international
competitiveness and social cohesion / edited by Maria João Rodrigues.
    p. cm.
  Includes bibliographical references and index.
      1. Technology—Economic aspects—European Union countries.
  2. Technology and state—European Union countries.   3. Education and
  state—European Union countries.   4. Knowledge, Sociology of—European
  Union countries.   5. Competition, International.   I. Rodrigues, Maria João.
  HC240.9.T4 N49   2002
  303.48'3'094—dc21   2001031531

ISBN 1 84064 719 1 (cased)
       1 84376 468 7 (paperback)

Printed and bound in Great Britain by MPG Books Ltd, Bodmin, Cornwall

Contents

Figures

Tables

Contributors

Maria João Rodrigues
Professor of Economics at the University of Lisbon – ISCTE
Chair of the Advisory Group of Social Sciences in the European
    Framework Programme for Research (EU)
Special Advisor to the Prime Minister

Luc Soete
Professor of International Economics at the Faculty of Economics,
    University of Maastricht
Director of the Maastricht Economic Research Institute on Innovation
    and Technology (MERIT)

Gøsta Esping-Andersen
Professor at the Universita di Trento
Professor at the Universitat Pompeu Fabra, Barcelona
Member of Scientific Council, Juan March Foundation

Robert M. Lindley
Professor, Faculty of Social Studies, University of Warwick
Director, Institute for Employment Research, University of Warwick
Associate Director of the Business Processes Resource Centre of the
    Economic and Social Research Council, based at the International
    Manufacturing Centre, University of Warwick.

Robert Boyer
Professor at the Ecole des Hautes Etudes en Sciences Sociales, Paris
Economist at CEPREMAP
Senior Researcher at CNRS
Head of URA CNRS 922 'Regulation, Human Resources and Economic Policy'

Bengt-Åke Lundvall
Professor in Economics at the Department for Business Studies at the
    Aalborg University
Research Manager for the nation-wide Danish network DRUID (Danish
    Research Unit for Industrial Dynamics)

Mark Tomlinson
Research Fellow at the Centre for Innovation and Competition,
    University of Manchester

Manuel Castells
Professor of Sociology and Professor of City & Regional Planning,
    University of California at Berkeley
Research Professor, Consejo Superior de Investigaciones Científicas,
    Barcelona (on leave)

Mario Telò
Professor, Université Libre de Bruxelles
J. Monnet Chair for Political Problems of the European Union
Research Director, Institute for European Studies (IEE, ULB)

Acknowledgements

This book presents the main outcomes of a prolonged and complex interaction between the scientific and the political agenda at the European level. Two kinds of outcome have emerged: new theoretical issues and a long-term strategy for the European Union. This interaction was made possible by a new kind of alliance between the intellectual community and the political community which respected their autonomy and was fostered by some mediators belonging to both, as in my case as coordinator of this process.

Therefore my first acknowledgements go to Prime Minister António Guterres who, from the outset, had the vision and the ambition to launch an extensive process for long-term thinking and decision, involving his colleagues, heads of state and prime ministers; enthusiastic contributions were received from all of them which, subsequently, gave the impetus to the hundreds of contributions which followed from political bodies and civil society in Europe. A very important contribution also came from the various members of the Portuguese government who 'spread the word' on behalf of the Presidency of the European Union. In our various tours of capitals we have nearly always received very positive feedback.

My second acknowledgements go to my colleagues, and now friends, authors of this book who understood the magnitude of the opportunity and of the challenge from the beginning and committed themselves to preparing their own contribution and to exposing their own ideas to a large and diversified public. I want to stress the privilege of having their collaboration, the outstanding memories of our talks, our e-mails and our meetings. The European research community is made of these stories.

I am specially grateful to Jacques Delors, a master of this interaction between politics and science, who gave me the privilege of some appointments at crucial moments. Otherwise experienced mediators played an important role throughout the process and among them I would particularly like to mention and thank Prime Minister Jean-Claude Juncker and Director General Allan Larsson. I also always kept my former scientific director in Sorbonne University, Henri Bartoli, as a main reference.

In the meantime, many other authors from different countries gave a very relevant contribution to this public debate, namely Jos Berghman, Giuseppe Bertola, Olivier Blanchard, Albert Bressand, Bernard Brunhes, Yves Chassard, Vítor Constâncio, Maurizio Ferrera, Jacques Freyssinet,

Stephen Fuller, Carlos Laranjo Medeiros, Lucio Pench, Maria de Lourdes Pintasilgo, Marino Regini, Michael Shuman, Spiros Simitis, Göran Therborn, Jürgen Von Hagen.

The same applies to many actors from the political community. At the European level, President Romano Prodi, the Commissioners Pedro Solbes, Erkki Liikanen, Anna Diamantopoulou, Philippe Busquin and Viviane Reding should have a very special mention along with other representatives of the European Commission, namely David O'Sullivan, Carlo Trojan and Michel Petite, Directors-General such as Allan Larsson, Robert Verrue, Fabio Colasanti, Domenico Lenarduzei, Odile Quintin and, of course, the President of the European Parliament, Nicole Fontaine, and the chairmen of the EP Committees Christa Randzio-Plath and Michel Rocard. Very active and important contributions also came from social partners' leaders, namely Emilio Gabaglio and Dirk Hudig. At the national level, I am particularly grateful to my colleagues in the Prime Ministers' Cabinets of all Member States, Kaare Barslev, Markus Beyrer, Hugo Brawers, Thomas de Bruijn, Stefan Collignon, Anna Ekström, Román Escolano, Gilles Gateau, Klaus Gretschmann, Gikas Hardouvelis, Jeremy Heywood, Philip Kelly, Roger Liddle, Jari Luoto, Florence Mangin, Jeppe Tranholm-Mikkelsen, Pierre-Alain Muet, Anders Nordström, Pier Carlo Padoan, Pentti Puoskari, Platon Tinios, Baudílio Tomé, Luc Wies.

Last but not least, I would like to express my sincere thanks to my operational team Patrícia Cadeiras, Rui Moura and Isabel Cernich, as well as to Rachel Evans, whose competence and stamina made this whole undertaking possible. Edward Elgar's prompt and positive reaction gave us the final and decisive impetus to prepare this book.

While all these references are due, in no way do they erase my responsibility as the editor and coordinator of this action line of the Presidency of the European Union.

I hope this publication will be useful for all who want to discuss and to build Europe both in political and scientific terms.

1. Introduction: for a European strategy at the turn of the century

Maria João Rodrigues

At the turn of the century, we must think in the long term. With this endeavour, the Presidency of the European Union organised a special interaction between the scientific and the political agenda. The purpose of this introduction is to reveal how this cross fertilisation was developed, leading to a new European strategy which aims to build a knowledge-based economy with more competitiveness and social cohesion. The main political dilemmas, the main theoretical issues, the new proposals and, finally, the political outcomes will be presented, providing a preliminary framework to highlight the following chapters.

1. EUROPEAN DILEMMAS

1.1. What scenarios for Europe?

Europe is facing a crucial period in its history. All the main issues of European civilisation are at stake and very contrasting scenarios are possible as shown in a wide range of literature (Delors, 1992; Wallace, 1990; Jacquemin and Wright, 1993; McRae, 1994; Bressard, 1997; Telò, 1998 and Fitoussi, 1999).

There is a bifurcation in each of the main factors shaping the European scenarios:

- In the international order, will we have a lasting American leadership or a more multipolar structure?
- In enlargement, will we have a slower or a faster pace?
- In the institutional reform, will we have a more confederal or a more federal evolution?
- In the creation of a single currency, will we have more or less credibility in the financial markets?

- In the specialisation pattern, will we have more or less polarisation between high-skilled and low-skilled European regions?
- In macroeconomic policy, will the emphasis of the fine tuning be on inflation or on unemployment?
- In structural policies, will we have less or more effectiveness in economic redeployment?
- In combating the different risks of social exclusion, will we have less or more effective policies?

The most likely combinations of these alternatives can lead us to some very different scenarios:

- In the scenario 'Slow integration', Europe retards both enlargement and deepening, faces difficulties in affirming euro, cannot avoid the rise of unemployment and social exclusion and loses influence in the international game.
- In the scenario 'Enlargement as the priority', Europe speeds up enlargement and reaches more credibility with euro, but faces some difficulties in regional development, employment and social exclusion.
- In the scenario 'Enlargement and deepening', Europe also has success in enlargement and in the single currency, achieving it with more economic and social cohesion associated with some kind of political deepening and increasing international influence.

Other combinations and other scenarios are of course possible and this makes the European path a complex and an uncertain one.

The focus of this book will be on the economic and social development of the European Union, whilst bearing in mind this more general framework.

1.2.   A new paradigm creating a new context

Europe is at the crossroads in a changing landscape. A completely new environment is being created by globalisation, technological change and an ageing population with its impact on the welfare state. With globalisation, nations are competing to attract investment, which, on the one hand, depends increasingly on the general conditions supporting business competitiveness. On the other hand, business competitiveness depends increasingly on the capacity to answer just in time to the specific needs of the customer. This involves managing a greater amount of knowledge with the intensive use of information technologies.

Knowledge is becoming the main source of wealth of nations, businesses and people, but it can also become the main source of inequalities among them. A new paradigm is emerging creating knowledge-based economies and societies. This is the broader significance we should give to the recent terminology about the 'new economy'.

Knowledge has always been an ingredient of human societies, but what is radically new is the speed of its accumulation and diffusion, due to information and telecommunication technologies. Working conditions and living conditions are being redefined. Markets and institutions are being redesigned by new rules based on the new possibilities of exchanging information. Internet is becoming the main infrastructure of this new paradigm.

Europe is somehow lagging behind in this transition and can learn a lot from the United States. But the point is not to imitate the United States, but rather to define the European way to the knowledge economy.

The challenges embodied in the European scenarios must be re-examined in the light of this emerging paradigm.

1.3. Dilemmas and possibilities

Some of these challenges concern competitiveness on the one hand and social cohesion on the other. Here we have a crucial dilemma. A realistic assessment might conclude that it is not possible to keep up with the so-called European social model, as this is now, in the new conditions created by globalisation and technological change, aggravated by ageing populations. Hence, a defensive answer to this prospect might consist of downgrading this European social model in order to increase competitiveness. A more affirmative answer, and also a more complex one is threefold: to build new competitive factors, to renew the European social model and to regulate globalisation.

Regulating globalisation depends on the ongoing reform of the United Nations and Bretton Woods institutions, namely the role of the International Monetary Fund (IMF) in financial markets, and on the next round of the World Trade Organisation (WTO) in order to foster multilateral trade. Better coordination of the foreign policies of European countries can also play a relevant role in this framework.

When building new competitive factors, a range of possibilities opened up by the knowledge-based economy should be explored in order to modernise companies, public services, schools, transports, cities and all the surrounding environment.

Renewing the European social model should create the conditions to help people move from jobs with no future to jobs with a future. This involves active employment policies, education and training, collective bargaining with a greater focus on change and more active social policies ensuring a safety network. This also involves making a special effort to prevent the digital divide, the new forms of social exclusion arising from the information society.

1.4.   Two central questions for Europe

Two central question seem to emerge for Europe: How is it possible to speed up the transition towards a knowledge-based economy with more jobs and more social cohesion? How is it possible to make Europe a more competitive and dynamic economy, able to create more and better jobs and greater social cohesion?

These were exactly the central questions posed to Europeans leaders at the beginning of the century, the right moment to think in the long term.

As a background to their decision making, they were given a broad picture on the emerging knowledge-based economy.

2.   THE EMERGENCE OF THE KNOWLEDGE-BASED ECONOMY AS A GREAT TRANSFORMATION

2.1   The nature of the knowledge-based economy

The knowledge-based economy is more than the so-called new economy. The fashionable term 'new economy' is sometimes limited to software and multimedia business, supported by active financial markets. But this is the tip of the iceberg. A much wider change is going on which encompasses all sectors of activity, from services, to manufacturing and even agriculture under the pervasive effect of information technologies and telecommunications (Cairncross, 1997; Thurow, 1999). A deluge of technological innovations is invading all these sectors and transforming our lives, from computers to computer-aided manufacturing, and from mobiles phones to digital TV, but even the other usual concept of 'information society' is limited to capturing the in-depth nature of the ongoing change.

As a matter of fact, the ongoing change is not only technological but also institutional, and it concerns something more than information, namely knowledge. We are living through a great transformation (with the meaning given by Polanyi [1944], 1983) which concerns the very social processes of knowledge production, diffusion and utilisation. Knowledge accumulation was speeded up in the past by major inventions, such as writing and printing. Communication between different communities was made more independent of their co-existence at the same time and in the same space.

The current technological revolution is making human communication even more independent of time and space constraints, speeding up knowledge accumulation. The available knowledge at the cultural or scientific level is transformed in new contents and widely spread by increasingly powerful combinations of software and hardware. The knowledge intensity of products and services is also increasing, as can be seen in transport, health, education or entertainment. Knowledge is becoming the main raw material in many manufacturing companies. All social institutions work in a different way and even markets become more knowledge intensive, as displayed by financial markets or e-commerce.

A virtual reality is being built, the so-called cyberspace, whose main rules and architecture are still being defined, but which is already having powerful interactions with the existing world (e.g., Lévy, 1997). The organisation of the cyberspace is re-organising the existing world, shaping its economic, financial, political and cultural exchanges. Cognitive capacities, connectivity and cultural identity become the key instruments for survival in the new world. Internet and its social use is the most striking outcome of all this great transformation.

2.2.   Economic and social implications of the knowledge-based economy

Knowledge is becoming the main source of wealth and power, but also of difference, between nations, regions, companies and people (Castells, 1996). Innovation based on a specific knowledge is the main competitive advantage. Competitiveness means to answer just in time to the personal needs of the customer, which requires a very sophisticated knowledge management. Mass customisation is succeeding to fordist standardised mass production (Tapscott, 1995). The foremost companies focus on the most value-added productions, build trade marks and launch wider operations of outsourcing and delocalisation. Network companies are spreading in all sectors and nations, reorganising the international division of labour. With e-commerce, businesses trade directly with businesses and the company dimension can become more irrelevant when taking advantage of globalisation. But soon the old intermediators are replaced by new intermediators capable of reorganising the market places in the cyberspace.

Knowledge management becomes a key component of corporate strategic management, activating the relationship between marketing, research and production. Corporate organisation is reshaped to build a learning organisation. New types of workers emerge, knowledge workers who have been categorised in different ways (Reich, 1991). Castells (1996) identifies new profiles, such as captains, innovators and connectors. Human resources management focus with increasing sophistication on the production of new competences as a source of competitiveness (e.g., Le Boterf, 1998). In the meantime, new risks of social exclusion, of a digital divide, emerge involving all the workers who cannot keep up with this pace of change.

Labour markets tend to new forms of segmentation between workers with voluntary mobility based on up-dated skills and workers who run the risk of involuntary mobility due to out-dated skills. The institutional framework of labour markets is being shaped in order to recombine employability and adaptability with basic conditions of security and citizenship (e.g., Esping-Andersen, 1999; Fitoussi and Rosanvallon, 1996). Labour market services are more focused on active employment policies; social protection systems on activating social policies; industrial relations on negotiating new trade-offs between flexibility and security. Finally, education and training systems are facing the challenge of building a learning society as a pre-condition to having a knowledge-based society and not only a knowledge-based economy (e.g., Lindley (with Nadel), 1998). To sum up, institutional innovations are emerging and new social rules are being invented.

How can we highlight our possible paths in this great transformation? In order to foresee and to discuss the possible scenarios, we must come back to some foundations underpinning this analysis of the emerging knowledge-based economy.

3.   THE INTELLECTUAL HORIZON AT THE TURN OF THE CENTURY

Thinking in the long term at the turn of the century requires a prospective effort, building on our intellectual legacy. Hence, some major breakthroughs of the past century should be underlined, namely in the philosophy of knowledge, the philosophy of science and the philosophy of politics, because they are shaping our intellectual horizon.

3.1.   Philosophy of knowledge: theory shaping empirical evidence

The approach about the relationship between theory and empirical evidence is a key issue for the development of science. The central controversy in the philosophy of science between rationalism and empiricism yielded important outcomes throughout the twentieth century. Overcoming the established tradition of the experimental method giving birth to hypotheses, the Vienna school stressed the preliminary role of theory by defining the hypothetico-deductive method. Moreover, the role of empirical evidence was more accurately defined by Popper with the concept of the falsifiability of a theory as opposed to the illusionary search of the verifiability of a given theory.

The role of theory in the construction of facts was highlighted in many different ways throughout the century:

- by Heisenberg's principle of indetermination;
- by Kuhn's concept of scientific revolutions based on new paradigms;
- by the identification of epistemological obstacles, such as teoricism and empiricism undertaken by Bachelard;
- by the theoretical developments on the role of language and of language games by Wittgenstein;
- by Habermas' theory on communicational action [1981] (1987);
- by the new insights coming from semiotics and cognitive sciences, with the pioneering work of Simon in economics and, very recently, from neurobiology, with Damásio (1995).

More recently, the transition of the century was signalled with a major transformation of the processes of knowledge accumulation and diffusion, hugely accelerated by information and telecommunication technologies. As the new cyberspace is being organised, it is becoming even more clear that there is no linear accumulation of knowledge. Instead there is creation and destruction of knowledge, conflicts for influence and emerging hegemonic poles with power to structure knowledge.

3.2. Philosophy of science: complexity and multidimensionality

Shifts in the philosophy of knowledge led to shifts in the philosophy of science. Cubist painting was perhaps the best pre-intuition of what was about to come. The object we see depends on the perspective. The complexity and the multidimensionality of the scientific object was also richly highlighted throughout the century, namely:

- in physics, by Einstein's theory of relativity;
- in biology, with the introduction of the concepts of evolution, the genetic code, irreversibility, entropy and regulation, as shown by Prigogine and Sengers;
- in human and social sciences with the introduction of the role of values, representations, social norms, institutions, as well as individual freedom in understanding, choosing and behaving, as recently re-elaborated by Morin and Naïr (1997), Sen (1999), Bartoli (1991), Bourdieu (1979) or Giddens (1984).

The scientific paradigm based on Newtonian physics, a reversible mechanism without time dimension, still survives in different sciences, including economics. Nevertheless it seems increasingly a simplification with too many limits, only useful in a small range of cases (e.g., Parrochia, 1997). Yet, science must be able to simplify complexity in order to explain, predict and transform. A new generation of models and mathematical tools is now being created, involving artificial intelligence in order to cope with complexity and multidimensionality.

One of the most outstanding implications of this new paradigm regards history and, more precisely, the philosophy of history. All social facts are embedded in history but there is neither a historicism, a pre-defined finality of history as stated by Marxist theory, nor an end of history as claimed by Fukuyama (1992). There are some strong trends with determination power, but there is also a degree of uncertainty and of influence coming from human action. This kind of approach can overcome the arguments raised by the post-modernist movement, very fashionable at the end of the last century.

Hence it is possible to speak not about a new historical stage but about a new historical situation emerging at the dawn of the century. The world order is being reshaped again, after a century of strong convulsions.

The twentieth century closes a long period in Europe marked by nation-states trying to create their own empire, with a final clash in the Second World War. But it is also a century marked by the experiments of the socialist revolution and subsequent developments of the Cold War, intertwined with decolonisation and the settlement of world-level organisations. Finally, the implosion of the Soviet Union and an enhanced hegemony of the United States co-exist with a faster diffusion of democracies and the emergence of regional blocs aiming at economic or even political integration, with a leading experience in Europe.

3.3. Political philosophy: globalisation, social justice and governance

The multiple implications of globalisation have became apparent more recently, not only in trade, industry and financial markets, but also in politics, culture and media (e.g., Held *et al.*, 1999). Global actors coming from civil society are the final demonstration. Heavy imbalances seem to emerge at social and environmental levels, calling for new forms of regulation at the world level (e.g., United Nations, 2000; Bartoli, 1999).

However, globalisation is also opening up important opportunities for growth and development, and radical technological innovations, like those in information technologies and genetic engineering, are creating powerful tools to answer these imbalances. As shown by Kuhn (1970), a new research programme usually arises from the confluence between new problems and new approaches. One typical research programme for the twenty-first century could be to focus on a central issue: how to use information and genetic technologies to reduce the social and the environmental imbalances at world level?

Nevertheless, the implementation of the possible solutions pointed out by this research programme would be mainly a political problem, requiring political choices.

In the field of political philosophy, the past century also leaves a legacy of remarkable redeployment. After a period of strong ideological cleavages, there seems now to be a greater convergence in the values of liberty, formal legal equality and, more recently, even in individual initiative and responsibility (e.g., Eatwell and Wright, 1993 and Giddens, 1998). Political controversy seems now to be more focused on defining social justice (see Rawls [1972], 1989) and setting the level and the forms of solidarity required to overcome the social imbalances.

New approaches are also being proposed in order to redefine political action and the relevant political structures. Bearing in mind all this new intellectual horizon highlighted above, one could ask: how to define a method of political action able to deal with complexity and uncertainty, to build on available knowledge, to create a larger consensus in civil society and to overcome the main imbalances? As a matter of fact, this is a central issue for an already ongoing action-research programme on governance. Governance is not only a matter of government but of many others actors, relevant to political action and regulation (e.g., Kooiman, 1993).

Taking into account the above-mentioned imbalances at the world level, some key problems for the present century will therefore be: how can we build a multilevel system of governance combining the world, the regional, the national and the sub-national level? How can this multilevel system of governance promote openness, diversity and cohesion at the world level?

3.4. Europe at the crossroads

Europe is at the crossroads of all these new issues of philosophy of knowledge, science and politics. Many reasons can be pointed out to explain this:

- first of all, because Europe was the first source of many of these theoretical and practical issues;
- because Europe carries a special responsibility in the new stage of knowledge production and diffusion created by cyberspace: what is at stake is to safeguard critical spirit, theoretical imagination, cultural and linguistic diversity. Multidimensionality and multiculturalism should be fostered, not impoverished.

- because European countries have lost the illusions of rebuilding the empire and have engaged themselves in an ambitious process of cooperation and integration;
- because European countries are exploring a new path for governance based on a multilevel system encompassing local, national, regional and even world levels;
- finally, because European countries, due to their history, provide a wide range of sensitivities and knowledge about the problems of third countries.

From these different points of view, Europe can be considered a vital laboratory in the search for new solutions to cope with the new issues identified above.

4.   NEW INSIGHTS IN SOCIAL SCIENCES

4.1.   Forecasting exercises and social theory

Keeping this impressionistic picture of our legacy as a background, let us now turn to the future. But some theoretical and methodological considerations should be presented before identifying some key questions for the definition of a European strategy.

Prospective exercises and scenarios building are today a burgeoning discipline, widely accepted by both academic and political communities and by society at large. They are a device to cope with complexity and uncertainty and to support strategic choices at different levels. After the pioneering works of Forrester, diversified exercises are having an impact on policy making, such as the reports produced by OECD Futures Programme (2000).

All the methodological sophistication can be found in books, such as those of Kees Van Der Heijden (1996) and of Michel Godet (1991). Different stages must be gone through in order to build scenarios: identifying the key variables and analysing their relationships; examining the strategies of the different relevant actors regarding these structural relationships; exploring the possible scenarios; reducing uncertainty by identifying those scenarios which are most probable and those which could be more desirable; finally, drawing some conclusions for strategic decision making and for policy making. Building scenarios has also proved to be a stimulating learning process for those involved in the exercise.

Nevertheless ongoing experience has also shown that the forecasting and strategic capacity of these methodological instruments depend on the explanatory strength of the theoretical framework previously adopted, given that it determines the choice of the relevant variables, relationships and actors. Empirical research is also very important afterwards in order to estimate the probability of each scenario.

The theoretical framework adopted in whatever social science, in order to build scenarios and make strategic choices, is based on assumptions of fundamental social theory, even if they are not explicit. As shown by Turner ([1974], 1991), the different currents we find in the history of social theory can be classified according to two main criteria: the level of theorisation and the drive for change. The level of theorisation can be either macro or micro and the drive for change can either be mainly conflict and imbalances or cooperation, equilibrium and homeostasis.

Hence, the theories which focus on macro-theorisation are divided between those which emphasise conflict, such as the theories of Marx, Weber, Habermas or Dahrendorf, and those which emphasise equilibrium, such as the theories of Spencer, Durkheim, Parsons or Luhmann.

On the other hand, the theories which focuss on micro-theorisation range from exchange and rational choice theories (Mauss, Hechter) to interactionism (Mead, Goffman) which is more open to different representations of the reality and different rationalities according to the actors.

Some theories have been built more recently to overcome these opposing positions, in order to embody both macro and micro approaches and both conflict and cooperation as drivers for change. The traditional controversy between holism and individualism are explicitly overcome by the theories developed by Giddens (1984), Bourdieu (1979) or Burns and Flan (1987). There is a permanent interaction between structure and actors and not only a one-way relationship.

Burns' (1987) approach is particularly interesting for economics, if we accept that market dynamics depend not only on prices but also on social rules. Burns studies the system of social rules which structure and regulate social transactions and organisations. He also studies the very processes of production, interpretation and implementation of these social rules, involving learning, negotiation, conflict and the exercise of power. A stimulating theoretical framework is provided to analyse bottlenecks, innovations, structural change, strategic action and reform, either at the macro or micro level.

How far is economic thought absorbing these new insights coming from social theory?

4.2. Economic dynamics and social context

A crucial dilemma seems to have marked economic thought over the past century. Should the social and historical context be taken into account in order to analyse economic dynamics or should it be virtually sacrificed in order to enable modelisation and more precise explanation and forecasting? At the turn of the century, serious attempts are being made in order to overcome this dilemma.

The influence of the social and historical context had explicitly been analysed by the institutional school with Veblen, or by the Austrian school with Schumpeter, but it was neglected in the major developments in the microeconomic foundations undertaken by Walras and Marshall. It was taken into account in Keynes' major breakthrough, aimed at building macroeconomic tools to analyse short-term fluctuations of growth and employment creation. But Keynesian theory faced some difficulties adapting itself to the new context of open economies, increasing globalisation, the role of rational expectations and the vanishing of monetary illusion. The criticisms coming from monetarists and the new classics explored these weaknesses of Keynesian theory, but in a field where the historical and social dimensions were deliberately excluded in order to build alternative theories with the main concern, not of realistic assumptions, but rather of operational instrumentality and falsifiability.

The importance of controlling inflation is much more consensual today even among Keynesian economists, but the different sensitivities between Keynesians and monetarists continue to have some relevance. These differences concern namely the impact of monetary or fiscal policy and the possibility of stimulating demand in order to foster growth and employment. Nowadays, however, this only requires a fine-tuning of the policy mix in the framework of sustainable growth.

These controversies also had the merit of pointing out the need to go beyond the policy mix options in the short term and to focus on the very foundations of economic growth, in the long term. Following the models of Harrod–Domar, Solow and Cobb–Douglas, a new generation of models based on endogenous growth highlighted the role of investment in research and development (R&D) and human capital in fostering growth and catching up (e.g., Romer, 1990, Silverberg and Soete, 1994). The same principle has been emphasised by neo-factorial and technological theories in international economics (e.g., Krugman and Obstfeld, 1994).

Reaching this frontier, economic theory seems to incorporate new dimensions of analysis, although it does not make them explicit. What is behind the investment in R&D and human capital, the increasing mastering of a concrete technology and a given technological trajectory? As shown by the evolutionary theory (e.g., Dosi *et al.*, 1988; Lundvall, 1992), we can find learning processes which occur in different institutional frameworks, with different rules and different actor games. Different social arrangements can underpin learning by doing, learning by using and learning by interacting.

Analysing long-and short-term dynamics by unveiling their institutional frameworks has been an area of convergence of more recent approaches, such as the evolutionary theory (Hodgson, 1993) and the regulation theory (Aglietta, 1976; Boyer and Saillard, 1995), which explains how the historical emergence of a new growth regime also depends on the renewal of social rules and institutional forms. These recent theoretical developments provide interesting tools with which to think about the role of structural policies and of institutional reforms in fostering growth and employment. More precisely, the emergence of a new growth regime based on knowledge seems to depend on institutional innovation, on the invention of new social rules.

Building on the legacy of the past century, a new theoretical paradigm also seems to emerge for social sciences based on the following principles:

- building a conceptual framework based on more realistic assumptions, able to take into account the diversity, complexity and multidimensionality of situations;
- going beyond the simplification of the homo economicus and taking into account different levels of information and different patterns of rationality;
- overcoming the limits of methodological individualism by accepting a permanent interaction between individuals and structures, institutions and social rules;
- admitting asymmetric relationships between actors which can involve either conflict or cooperation;
- taking not the equilibrium conditions but the evolution, the innovation and the learning conditions as the points of departure for analysis, even if the final purpose is also to explain equilibrium.

At the dawn of the century we are confronted with a crucial doubt: how far can a new theoretical paradigm enable us to cope with new economic and social paradigms?

Hence, preparing for the transition to a knowledge-based economy and society seems to be a difficult challenge both in theoretical and political terms. To cope with this challenge, cross-fertilisation between the scientific and the political agenda was put under way at the European level.

5.   RESHAPING THE EUROPEAN AGENDA

5.1.   Our main issues

Despite a number of undeniable successes, Europe is lagging behind in this transition to an innovation- and knowledge-based economy. This delay is apparent in the production and dissemination of many information technologies, but also in the adaptation of social institutions and social relations to the new growth potential opened up by these technologies. While this failure to adapt to the new paradigm continues, there will be a shortfall in economic growth and an increased risk of unemployment and social exclusion.

In order to foster this growth potential, institutional reform should focus on innovation systems, R&D systems, financial systems, education systems, labour market management and social protection systems.

Moreover, this institutional reform has not only a national but also a European dimension. National realities are very diversified in those various systems (Albert, 1991) but they face common problems of structural adjustment and they are influenced by a common framework based on a single market, a single currency, common competition and monetary policies, coordinated macroeconomic policies and some common standards at social and environmental levels.

This means that some level of European coordination is required in order to undertake these institutional reforms, while respecting national specificity. A multilevel governance system is needed based on the interaction at the European, the national and the sub-national levels.

All these issues led to a central question being analysed when preparing for the Presidency of the European Union: in order to strengthen its growth perspectives and to sustain its basic social model, Europe will have to build a new competitive platform based on innovation and knowledge. But there are risks in this transition. How can a knowledge-based economy be built with more and better jobs and more social cohesion?

Against this background, some more precise questions were identified to be addressed with intellectual autonomy by the different authors of this book, well known for their approaches and important contributions in these fields. In particular:

- by Luc Soete, the challenges and the potential for growth of a knowledge-based economy in a globalised world;
- by Gøsta Esping-Anderson, the new challenges for the welfare state resulting from an ageing population, globalisation and the emergence of a knowledge-based economy;

- by Robert Lindley, the emerging trends in employment, the need for new skills and the new threats of social exclusion in the transition to a knowledge-based economy;
- by Robert Boyer, the institutional reforms required to implement a policy mix for growth, employment and social cohesion in the transition to a knowledge-based economy;
- by Bengt-Åke Lundvall, the role of international benchmarking as a policy learning tool;
- by Manuel Castells, the construction of the European identity against the background of these societal changes;
- by Mario Telò, the building of a multilevel governance with government required to undertake these reforms.

The main purpose was to identify which institutional innovations could change the way in which European societies are currently regulated, so as to pave the way for a new development trajectory, for a more desirable scenario among the possible ones. This required the in-depth organisation of interactions and a kind of cross-fertilisation between the scientific and the ongoing European political agenda. The outcome was supposed to be a review of the European agenda in the light of this new set of issues.

5.2. The topics in discussion in the European agenda

The Portuguese Presidency of the European Union at the beginning of 2000 endeavoured to make a contribution to the long-term decision to update the European strategy for growth, competitiveness and employment in the light of the new conditions and building on previous outstanding contributions (namely, European Commission, 1994). This strategy should comprise macroeconomic policies, economic reforms and structural policies, such as R&D, innovation and enterprise policies, education policy, employment policy and social protection policy.

In the preparatory period of this Presidency, some main issues were being discussed at the European level based on a long list of documents.

Regarding macroeconomic policy:

- How to improve the policy mix for sustainable growth and employment in the framework defined by the Stability and Growth Pact?
- How can monetary policy contribute to growth if its main task, defined by the Treaties, is to control inflation?
- How far can automatic stabilisers of budgetary policy be used in order to sustain growth?
- What should the priorities of public investment be?

- How far should tax policy go in coordination?
- How should wage developments be made compatible with inflation control and rising living standards?
- What should the role of the macroeconomic dialogue with social partners be in the framework of the Cologne process?
- How can the risks of social dumping be avoided?
- How can the effectiveness of the official document of Broad Economic Policy Guidelines (BEPG) be improved?

Regarding the coordination between macroeconomic policies:

- What should the economic coordination be between member states: setting common objectives, monitoring their implementation at the national level as already done? Or to go further aiming at an evaluation of the aggregate effects of the different national choices, or even at an implementation of common tools to deal with problems, such as asymmetric or global shocks?

Regarding economic reforms:

- Which should the priorities for economic reform be in the fields of liberalisation, privatisation, competition policy, public services modernisation, financial markets integration, venture capital development, support to small and medium-size economies (SMEs)?
- How should the effectiveness of the Cardiff process of economic reforms be improved?

Regarding employment policies:

- How should the high unemployment rate in Europe be reduced? Should it be possible to adopt a common target about the unemployment rate? How should an employment strategy based on the Luxembourg process be developed in order to foster more and better jobs?
- How should the four pillars of employability, enterpreneurship, adaptability and equal opportunities be combined in order to renew the European social model?
- How should social partners and other actors be involved in this strategy?
- How should corporate social responsibility be encouraged?

- How should the increasing skill shortages be faced?
- How should lifelong learning be developed?
- What can the contribution of education policies be at the European level if they are only at a stage of cooperation?

Regarding social protection:

- What can a concerted strategy for modernising social protection be if these policies are at an even more preliminary stage of cooperation at the European level? Some priorities have been defined: making work pay and ensuring a stable income; enhancing the sustainability of pension systems; promoting social inclusion and providing a high-quality health system. But how should these priorities be implemented?
- What can the future of the welfare state in Europe be?

Regarding governance:

- How can strategic leadership be enhanced for better coordination between economic, technological and social policies in Europe?

5.3. A cluster of new political orientations

A cluster of new political orientations began to emerge from all the inter-action organised between the political and the scientific community in order to reshape the European agenda:

- defining a global strategy for a knowledge-based economy with social inclusion, with implications for the various economic and social policies;
- discussing not only the quantity but also the quality of public finances; giving a clear priority to public investments in R&D, innovation, education and training; using tax policy to support innovative SMEs; enhancing public–private partnerships to launch ambitious info-structures;
- improving macroeconomic coordination with a more clear definition of the role of each actor and each level of decision making;
- focusing economic reform on enhancing the potential for growth and innovation;
- launching an ambitious plan for the information society facilitating access to the Internet, stimulating new contents, combating info-exclusion, focusing new citizen needs in education, health, transports, environment;

- coordinating national policies in order to enhance European potential in R&D;
- going beyond the traditional reforms of education and training systems and exploring new ways of building a learning society with opportunities for all;
- reforming the European social model by investing in people, activating the welfare state and enhancing the fight against social exclusion in its old and new forms, such as technological illiteracy;
- re-thinking the nature and the scope of the welfare state, considering not only guaranteed income, but also personal services, quality of work and living opportunities as basic components of social cohesion and social justice;
- enhancing the sustainability of pension systems by raising the employment rate which requires namely to foster jobs creation in services, to improve employability, to reconsider early retirements and to enhance equal opportunities;
- examining the interaction between the European and the national forms of regulation;
- analysing the sources of European identity and the role of a multilevel system of governance to build it;
- improving the role of social dialogue and other forms of partnership in order to manage change;
- introducing innovations in political method in order to improve coordination, monitoring, accountability and learning processes. In short, a more knowledge-based policy making.

The first outcome of this cross-fertilisation between the scientific and the political agendas was the Document from the Presidency presented in an annex under the title 'Employment, economic reforms and social cohesion – towards a Europe based on innovation and knowledge'. The main scientific inputs are the core of this book. This political and scientific starting point would be followed by dozens of documents prepared by the European Commission, the Council, all European governments, the European Parliament and the other European institutions, the social partners and many others actors at European and national levels.

The Lisbon European Council, held in March 2000, defined and put into practice the European strategy for the knowledge-based economy – now formally recognised as the 'Lisbon Strategy'. Its results began to be felt in various policy areas, namely the information society, in research, innovation, the internal market, in education, employment, social protection, social inclusion – even in macroeconomic policies.

There is a story behind all this that deserves to be told, so that we can understand the extent of this strategy and the new set of problems and of solutions for Europe.

5.4. Tailoring and leading political change

However, key ideas needed to lead to political decision taking and action. The entire Presidency was tailored to achieving this goal, throughout its two European Councils, fourteen Councils of Ministers, an International Hearing (1999), seven Ministerial Conferences, and several sessions of the European Parliament and a high-level Forum grouping the major stakeholders in Europe and the member states.

As the main objective was to define a global strategy, the key role had to be played by the European Council – in synergy with the initiatives of the European Commission. The meeting of the European Council had to be special and focused only on this objective. We had to hold it sufficiently early to provide guidance for the following Councils of Ministers and sufficiently late to allow for the strong effort in persuasion required to reach agreement. This action relied on a series of ambitious initiatives formally proposed by the Presidency, at its own risk, resulting in multiple contacts made with all community bodies and national governments. Ultimately it led to the prime minister's visit to all EU capitals. Public debate also made it possible to collect a widely diversified set of contributions from civil society, from all EU governments and from all community bodies.

Decisions made at the Lisbon European Council helped define the final shape of the high-level consensus and mobilisation obtained during this process, by establishing more precise objectives, calendars and methods and defining the mandates of all Councils of Ministers involved. This propeller enabled the last meeting of the European Council at Feira to produce a set of concrete results, which began to be translated at the national level and developed during the following Presidencies.

5.5. The Lisbon Strategy

A new strategic goal and an overall strategy was defined by Lisbon European Council in 2000. Quoting its own Conclusions:

'The Union has today set itself a **new strategic goal** for the next decade: to become the most competitive and dynamic knowledge-based economy in the world capable of sustainable economic growth with more and better jobs and greater social cohesion. Achieving this goal requires an **overall strategy** aimed at:

- preparing the transition to a knowledge-based economy and society by better policies for the information society and R&D, as well as by stepping up the process of structural reform for competitiveness and innovation and by *completing the internal market;*
- modernising the European social model, investing in people and combating social exclusion;
- sustaining the healthy economic outlook and favourable growth prospects by applying an appropriate macro-economic policy mix.'

Lisbon Strategy set the following main political orientations:

- a policy for the information society aimed at improving the citizens' standards of living, with concrete applications in the fields of education, public services, electronic commerce, health and urban management; a new impetus to spread information technologies in companies, namely e-commerce and knowledge-based management tools; an ambition to deploy advanced telecommunications networks and democratise the access to the Internet, on the one hand, and produce contents that add value to Europe's cultural and scientific heritage, on the other;
- an R&D policy whereby the existing community programme and the national policies converge into a European area of research by networking R&D programmes and institutions. A strong priority for innovation policies and the creation of a European patent;
- an enterprise policy going beyond the existing community programme, combining it with a coordination of national policies in order to create better conditions for entrepreneurship – namely administrative simplification, access to venture capital or manager training;
- economic reforms that target the creation of growth and innovation potential, improve financial markets to support new investments, and complete Europe's internal market by liberalising the basic sectors while respecting the public service inherent to the European model;
- macroeconomic policies which, in addition to keeping the existing macroeconomic stability, vitalise growth, employment and structural change, using budgetary and tax policies to foster education, training, research and innovation;
- a renewed European social model relying on three key drivers, i.e. making more investment in people, activating social policies and strengthening action against old and new forms of social exclusion;
- new priorities defined for national education policies, i.e. turning schools into open learning centres, providing support to each and

every population group, using the Internet and multimedia; in addition, Europe should adopt a framework of new basic skills and create a European diploma to embattle computer illiteracy;

- active employment policies intensified with the aim of making life-long training generally available and expanding employment in services (especially care services) as a significant source of job creation, improvement of the standards of living and promotion of equal opportunities for women and men. Raising Europe's employment rate was adopted as a key target in order to reduce the unemployment rate and to consolidate the sustainability of the social protection systems;
- an organised process of cooperation between the member states to modernise social protection, identifying reforms to answer common problems, such as matching pension systems with population ageing;
- national plans under preparation to take action against social exclusion in each and every dimension of the problem (including education, health, housing) and meeting the requirements of target groups specific to each national situation;
- improved social dialogue in managing change and setting up of various forms of partnership with civil society, including the dissemination of best practices of companies with higher social responsibility.

5.6. Strategy and governance

The actual implementation of any strategy requires a political engine, i.e. a governance centre at the European level with the power to coordinate policies and adapt them to each national context. The Lisbon decisions made this governance centre stronger, in three ways:

- firstly, the European Council will play a stronger role as coordinator, henceforth devoting its spring meeting to the monitoring of this strategy, based on a synthesis report presented by the European Commission;
- secondly, the broad economic policy guidelines will improve the synergy between macroeconomic policies, structural policies and employment policy;
- thirdly, the Union adopted an open method coordination, which will begin to be applied to all policy fields, stepping up the translation of European priorities into national policies.

This method combines European coherence and respect for national diversity. It defines the required European guidelines in each policy domain, subsequently identifying best practices and reference indicators and, finally, materialising in national plans consisting of concrete targets and measures fitting each country's case. Its purpose is to set up a vast process of innovation, learning and emulation between European countries, in which the European Commission may play a new role as catalyst. The method actually aims to speed up real convergence, now that nominal convergence is being achieved in order to prepare for the single currency.

The political construction of Europe is being based on different political methods in accordance with the problems to be solved. Various methods have been worked out which are placed somewhere between pure integration and straightforward cooperation. Hence:

- Monetary policy is a single policy within the euro zone.
- National budgetary policies are coordinated at the European level on the basis of strictly predefined criteria.
- Employment policies are coordinated at the European level on the basis of guidelines and certain indicators, allowing some room for adjustment at the national level.
- A process of cooperation is beginning with a view to the modernisation of social protection policies, with due regard for national differences.

Policies aimed at building the single market, such as monetary policy or competition policy, are, logically, based on a stricter method of coordination in relation to the principles to be observed. However, there are other policies which concentrate more on creating new skills and capacities for making use of this market and responding to structural changes. They involve learning more quickly and discovering appropriate solutions. Such policies have resulted in the formulation of a coordination method which is more open to national diversity.

5.7. The open method of coordination

The open method of coordination is based on different stages:

- defining guidelines for the Union combined with specific timetables for achieving the goals which they set in the short, medium and long terms;
- exchanging best practices and, where appropriate, establishing quantitative and qualitative indicators and benchmarks against the best in the world and tailored to the needs of different member states;

- translating these European guidelines into national and regional policies by setting specific targets and adopting measures;
- periodic monitoring, evaluation and peer review organised as mutual learning processes.

The open method of coordination is a concrete way of developing modern governance using the principle of subsidiarity. This method can foster convergence in common interest and in some agreed common priorities while respecting national and regional diversities. It is an inclusive method for deepening European construction.

The purpose of the open method of coordination is not to define a general ranking of member states in each policy, but rather to organise a learning process at the European level in order to stimulate exchange and the emulation of best practices as well as to help member states improve their own national policies.

The open method of coordination is to be combined with the other available methods depending on the problem to be addressed. These methods can range from integration and harmonisation, to cooperation. The open method of coordination itself takes an intermediate position in this range of different methods. It is an instrument to be added to a more general set of instruments.

The European Commission can play a crucial role as a catalyst in the different stages of the open method of coordination namely by: presenting proposals on European guidelines, organising the exchange of best practices, presenting proposals on indicators, and supporting monitoring and peer review.

Monitoring and evaluation should be based on systemic approaches in the national context and should help to create a culture of strategic management and of learning by experience, involving all relevant partners.

The main source of inspiration for the open method of coordination is the Luxembourg process regarding European employment strategy. Following the Lisbon Summit conclusions, this method is now being implemented in different policy fields, namely, the information society, R&D, enterprises, economic reforms, education and social inclusion. An empirical and flexible approach is being used in order to develop and to adapt this method to the specific features of each policy field.

Developing the knowledge-based economy with social cohesion and promoting real convergence in Europe, by matching the community drive with national policies – this will be the main test for the Lisbon Strategy over the coming years. This challenge involves various complex issues which will be developed in the following chapters.

REFERENCES

Aglietta, Michel (1976), *Régulation et crises du capitalisme*, reprinted (1997), Paris: Odile Jacob.

Albert, Michel (1991), *Capitalisme contre capitalisme*, Paris: Seuil.

Artis, Michael J. and Norman Lee (eds) (1997), *The Economics of the European Union – Policy and Analysis*, Oxford: Oxford University Press.

Baldwin, Robert and Martin Cave (1999), *Understanding Regulation*, Oxford: Oxford University Press.

Bartoli, Henri (1991), *L'économie multidimensionnelle*, Paris: Economica.

Bartoli, Henri (1999), *Repenser le Développement – En finir avec la pauvreté*, Paris: UNESCO – Economica.

Blaug, Mark (1962), *Economic theory in retrospect*, reprinted in Cambridge: Cambridge University Press (1988).

Bourdieu, Pierre (1979), *La distinction – Critique sociale du jugement de gout*, Paris: Les Éditions de minuit.

Boyer, Robert (1986), *La théorie de la régulation: une analyse critique*, Paris: Editions La Découverte.

Boyer, Robert and Yves Saillard (1995), *Théorie de la régulation – L'état des savoirs*, Paris: Editions La Découverte.

Bressard, Albert (1997), *Europe 2012, Globalisation et Cohésion sociale*, Paris: Economica

Burns, Tom R. and Helena Flam (1987), *The Shaping of Social Organisation – Social Rule System Theory with Applications*, London: Sage.

Cairncross, Frances (1997), *The Death of Distance – How the Communications Revolution will Change our Lives*, London: Orion Publishing Group.

Castells, Manuel (1996), *The Information Age: Economy, Society and Culture – The Rise of the Network Society*, vol. I, reprinted 1999, Oxford: Blackwell Publishers.

Commissariat Général du Plan (Rapport du groupe présidé par Jean-Louis Quermonne) (1999), *L'Union européenne en quête d'institutions légitimes et efficaces*, Paris: La Documentation française.

Conseil d'Analyse Economique (1998), *Coordination européenne des politiques économiques*, Paris: La Documentation française.

Damásio, António R. (1995), *O erro de Descartes – emoção, razão e cérebro humano*, Lisbon: Publicações Europa-América.

Delors, Jacques (1992), *Le nouveau concert européen*, Paris: Odile Jacob.

Delors, Jacques (1994), *L'unité d'un homme*, Paris: Odile Jacob.

De Schoutheete, Philippe (1997), *Une Europe pour tous*, Paris: Odile Jacob.

Dosi, Giovanni, Christopher Freeman, Richard Nelson, Gerald Silverberg and Luc Soete (eds) (1988), *Technical Change and Economic Theory*, London: Pinter.

Eatwell, Roger and Anthony Wright (eds) (1993), *Contemporary Political Ideologies*, reprinted (1994), London: Pinter.

Edwards, Geoffrey and David Spence (eds) (1994), *The European Commission*, reprinted (1997), London: Catermill Publishing.

Esping-Andersen, Gøsta (ed.) (1996), *Welfare States in Transition – National Adaptations in Global Economics*, reprinted 1998, London: Sage.

Esping-Andersen, Gøsta (1999), *Social Foundations of Postindustrial Economies*, Oxford: Oxford University Press.

European Commission (1994), *Growth, Competitiveness, Employment – The*

Challenges and Ways Forward into the 21st Century, Luxembourg: Office for Official Publications of the European Communities.

Fitoussi, Jean-Paul (1995), *Le débat interdit – Monnaie, Europe, Pauvreté*, Paris: Arléa.

Fitoussi, Jean-Paul and Pierre Rosanvallon (1996), *Le nouvel âge des inégalités*, Paris: Seuil.

Fitoussi, Jean-Paul (sous la direction de) (1999), *Rapport sur l'Etat de l'Union européenne*, Paris: Fayard and les Presses de Sciences Po.

Fondation Alphonse Weicker (1997), *Europe 2012 – Globalisation et cohésion sociale: les scénarios luxembourgeois*, Paris: Economica.

Fries, Fabrice (1995), *Les grands débats européens*, Paris: Seuil.

Fukuyama, Francis (1992), *The End of History and the Last Man*, New York: Free Press.

Gaspard, Michel (1997), *Réinventer la croissance – Les chemins de l'emploi en Europe*, Paris: Editions La Découverte & Syros.

Giddens, Anthony (1984), *The Constitution of Society*, Cambridge: Polity Press.

Giddens, Anthony (1998), *The Third Way – The Renewal of Social Democracy*, reprinted (1999), London: Polity Press.

Godet, Michel (1991), *De l'anticipation à l'action – Manuel de prospective et de stratégie*, Paris: Dunod.

Godino, Roger (1997), *Les sept piliers de la réforme*, Paris: Albin Michel.

Habermas, Jürgen (1981), *Theorie des kommunikativen Handels*, reprinted (1985), Frankfurt: Suhrkamp Verlag.

Habermas, Jürgen (1987), *Théorie de l'agir communicationnel – Tome 2 Pour une critique de la raison fonctionnaliste*, Paris: Fayard.

Hatem, Fabrice (1993), *La Prospective – Pratiques et Méthodes*, Paris: Economica.

Hayes-Renshaw, Fiona and Helen Wallace (1997), *The Council of Ministers*, London: Macmillan.

Held, David, Anthony McGrew, David Goldblatt and Jonathan Perraton (1999), *Global Transformations – Politics, Economics and Culture*, Stanford, CA: Stanford University Press.

Hix, Simon (1999), *The Political System of the European Union*, London: Macmillan.

Hodgson, Geoffrey M. (1993), *Economics and Evolution – Bringing Life Back into Economics*, Cambridge: Polity Press.

Jacquemin, Alexis and David Wright (eds) (1993), *The European challenges post-1992 – Shaping Factors, Shaping Actors*, Aldershot: Edward Elgar.

Kelund, E., R. Boyer and R.F. Hébert (1990), *A History of Economic Theory and Method*, New York: McGraw-Hill.

Kooiman, Jan (ed.) (1993), *Modern Governance – New Government – Society Interactions*, London: Sage.

Krugman, Paul R. and Maurice Obstfeld (1994), *International economics: Theory and Policy*, New York: HarperCollins College Publishers.

Kuhn, Thomas (1970), *The Structure of Scientific Revolutions*, Chicago, CO: University of Chicago Press.

Lafay, Gérard (1997), *L'Euro contre l'Europe? – Guide du citoyen face à la monnaie unique*, Paris: Arléa.

Le Boterf, Guy (1998), *L'Ingénierie des compétences,* Paris, Éditions d'Organisation.

Lévy, Pierre (1997), *Cyberculture*, Paris: Editions Odile Jacob/Editions du Conseil de l'Europe.

Lindley, Robert (with H. Nadel) (eds) (1998), *Les Relations Sociales en Europe*, Paris: L'Harmattan.

Lundvall, Bengt-Åke (ed.) (1992), *National Systems of Innovation – Towards a Theory of Innovation and Interactive Learning*, London: Pinter.

McRae, Hamish (1994), *The World in 2020 – Power, Culture and Prosperity: A Vision of the Future*, London: HarperCollins.

Morin, Edgar (1987), *Penser l'Europe*, Paris: Gallimard.

Morin, Edgar and Sami Naïr (1997), *Une politique de civilisation*, Paris: Arléa.

OECD (2000), *The Creative Society of the 21st Century*, Paris: OECD.

Olivi, Bino (1998), *L'Europe difficile – Histoire politique de la Communauté européenne*, Paris: Gallimard.

Parrochia, Daniel (1997), *Sciences exactes et sciences de l'homme: les grandes étapes*, Paris: Ellipses.

Polanyi, Karl (1944), *The Great Transformation,* reprinted (1980), Boston, CO: Beacon Press.

Polanyi, Karl (1944), *La Grande Transformation aux origines politiques et économiques de notre temps*, reprinted (1983), Paris: Gallimard.

Popper, Karl (1959), *The Logic of Scientific Discovery*, New York: Harper Torch Books.

Porter, Michael E. (1990), *The Competitive Advantage of Nations*, London and Basingstoke: Macmillan.

Portuguese Presidency of the European Union, (1999), *International Hearing* (Giuseppe Bertola, Olivier Blanchard, Robert Boyer, Albert Bressand, Vítor Constâncio, Gøsta Esping-Andersen, Stephen Fuller, Robert Lindley, Bengt-Åke Lundvall, Carlos Medeiros, Lucio Pench, Marino Regini, Maria João Rodrigues, Michael Shuman, Luc Soete, Mario Telò, Göran Therborn, Jürgen Von Hagen).

Prigogine, Ilya and Isabelle Stengers (1979, 1986), *La nouvelle alliance*, Paris: Gallimard.

Prodi, Romano (1999), *Un'idea dell'Europa – Il valore Europa per modernizzare l'Italia*, Bologna: Società Editrice il Mulino.

Rawls, John (1972), *A Theory of Justice*, reprinted (1989), Oxford: Oxford University Press.

Reich, Robert B. (1991), *The Work of Nations,* Vintage Books.

Rocard, Michel (1996), *Les moyens d'en sortir*, Paris: Seuil.

Rodrigues, Maria João (1988, 1992), *O Sistema de Emprego em Portugal*, Lisbon: Publicações Dom Quixote.

Rodrigues, Maria João (1991, 1998), *Competitividade e Recursos Humanos – dilemas de Portugal na construção europeia*, Lisbon: Publicações Dom Quixote.

Rodrigues, Maria João (coord.) (2000), *Para uma Europa da Inovação e do Conhecimento – Emprego, Reformas Económicas e Coesão Social,* Lisbon: Celta.

Romer, Paul (1990), 'Endogenous Technological change', *Journal of Political Economy*, 98, 71–102.

Ross, George (1995), *Jacques Delors and European Integration*, Cambridge: Polity Press.

Sen, Amartya (1999), *Development as Freedom*, New York: Anchor Books.

Shapiro, Carl and Hal R. Varian (1999), *Information Rules – a Strategic Guide to the Network Economy*, Boston, MA: Harvard Business School Press.

Silverberg, Gerald and Luc Soete (eds) (1994), *The Economics of Growth and Technical Change – Technologies, Nations, Agents*, Aldershot: Edward Elgar.

Tapscott, Don (1995), *The Digital Economy: Promise and Peril in the Age of Networked Intelligence*, New York: McGraw-Hill.

Telò, Mario and Paul Magnette (1998), *De Maastricht à Amsterdam – L'Europe et son nouveau Traité*, Brussels: Editions Complexe.

Terestchenko, Michel (1994), *Philosophie politique – 1. Individu et société*, Paris: Hachette.

Thurow, Lester C. (1999), *Building Wealth – The New Rules for Individuals, Companies, and Nations in a Knowledge-Based Economy*, New York: HarperCollins.

Turner, Jonathan H. (1974), *The Structure of Sociological Theory*, reprinted (1991), Belmont: Wadsworth Publishing Company.

United Nations, Secretary-General (2000), *We Peoples – United Nations 21st Century*, New York: UN.

Van Der Heijden, Kees (1996), *Scenarios: The Art of Strategic Conversation*, Chichester: John Wiley.

Wallace, William (ed.) (1990), *The dynamics of European integration*, London: Pinter Publishers for the Royal Institute of International Affairs.

Walliser, Bernard and Charles Prou (1988), *La Science Economique*, Paris: Seuil.

Walliser, Bernard (1994), *L'Intelligence de l'Economie*, Paris: Editions Odile Jacob.

2. The challenges and the potential of the knowledge-based economy in a globalised world

Luc Soete

INTRODUCTION

The Portuguese presidency of the EU came at a crucial time. The first six months of 2000, the last year of the second millennium, represented to some extent an ideal moment to take stock of where Europe had been heading since the economic integration process had been set in motion after the massive devastations of the Second World War, and, in particular, how Europe had adjusted to the major, world-wide structural transformations. The last decade of the twentieth century has been a period of major structural transformations world wide, but also quite explicitly in Europe. At the beginning of the decade, one witnessed the albeit sudden collapse of the former communist East European countries and their rapid opening up to market-led economic incentives, with as the most extreme case the economic and political integration of East Germany with West Germany and the European Union (EU). A year later, the fifteen EU member countries formally entered the European Single Market: a process of economic integration still incomplete today in many, non-manufacturing utilities and service sectors but having nevertheless brought about a gradual opening-up of many, traditionally closed, domestic markets. Less precise in timing but again, once initiated, progressing at an accelerating rate, financial markets underwent during the nineties a dramatic, world-wide deregulation. Independent domestic monetary policy became something of the past. More recently, the European telecommunications sector became deregulated and liberalised. The resulting growth and variety in telecom services being offered has been so strong that at least up to 2000, and against most forecast and predictions, no overall job losses in the telecom sector took place, rather the contrary. And, finally, there was of course the macroeconomic convergence process leading to monetary union with the formal introduction of the euro on 1 January 1999 in eleven EU member countries.

Alongside these major, policy-led structural transformations, sometimes causing them, sometimes enabling them, there was of course the rapid rate of technological change in information and communication technologies. Three specific features of these technologies have been instrumental in bringing about further structural transformations in the economic, social and organisational framework of our society, opening up an increasing number of sectors to international trade and restructuring. First, the dramatic reduction in the costs of information and communication processing; second, the technologically-driven 'digital convergence' between communication and computer technology; and, third, the rapid growth in international electronic networking.

Many of these structural transformation processes, such as those with respect to financial markets or even the telecom liberalisation process have been global in nature. They have, however, involved much more some regions or areas than others. While Europe appears to have been at the centre of most of the structural transformations listed above, it seems, as yet, to have benefited least from the growth opportunities behind these structural transformation processes. In saying this, care must of course be taken in acknowledging the variety of experiences in individual member countries.

This is why we start our analysis with a very short, bird's eye overview of Europe's aggregate growth performance over the last decade(s): the EU as a whole in comparison with the US and Japan, and the individual growth performance of individual member countries. In a second section, we then address, albeit briefly, some of the main concepts used in the emerging so-called knowledge-based economy. Obviously the impact of new information and communication technologies does not limit itself to the manufacturing and distribution of goods and services. Similar, fundamental transformations are likely to occur in the production, distribution and organisation of research activities and knowledge generation more generally. In a third section we briefly discuss Europe's performance in knowledge investment. The evidence presented suggests that a significant gap has emerged over the 1990s between the US and Europe in knowledge investment, and information and communication technology (ICT)-related knowledge investment in particular. From this perspective the US seems to have been much more successful than Europe in making its transition in the 1990s towards a knowledge-based economy. In the fourth section, we draw some policy insights from this US experience. In particular, we highlight a number of features, which seem to lie behind the US growth story. The focus of our analysis is on those elements which might be of particular relevance to Europe and might lend themselves to concerted policy action at the European level. Again, it must be emphasised that the variety of individual European country experiences renders any such aggregate comparison difficult. Sometimes comparison with the US might even not be particularly relevant and more could be learnt from the 'best practice' experience of other EU member countries. However, as an aggregate large 'single' market, comparisons of the EU with the US are particularly revealing in highlighting the many remaining barriers and impediments to benefiting in the areas of knowledge investment and growth from Europe's large market.

It is from this perspective, that, as we stress in the conclusions, a number of policy recommendations can be formulated. Far from being complete and all encompassing, these recommendations illustrate that in a number of areas there is today a need for a new, post-EMU, knowledge-investment-based, European policy agenda.[1]

2. EUROPE'S RECENT GROWTH EXPERIENCE IN HISTORICAL PERSPECTIVE

We start the analysis with a brief overview of some of the main aggregate economic trends over the post-war period and the 1990s in particular. We focus on the European Union as a whole in comparison with the United States and Japan.

Figure 2.1 shows the growth rate in real gross domestic product (GDP) in the US, the EU and Japan. To facilitate comparisons, GDP levels have been set for 1991 at 1. From Figure 2.1 it emerges that since 1991, US real GDP has grown by some 38 per cent, an average of more than 3 per cent a year, the EU by some 19 per cent, an average of some 2 per cent a year and Japan by some 9 per cent, an average of less than 1 per cent a year. At the same time, as illustrated in Figure 2.2, employment grew also much more significantly in the US than in Europe or Japan. In Europe employment growth only turned positive since 1995, when output growth accelerated.

Behind this European average over the 1990s of 'jobless' growth, very different growth and employment trends can be observed for individual member countries. At one extreme, Ireland, the Netherlands and Denmark resembled much more the US pattern of rapid output and employment growth, whereas at the other extreme Germany and Italy witnessed sluggish output growth with declining employment. Rather than talking about an aggregate EU pattern of 'jobless' growth, it seems more appropriate to talk about a pattern of insufficient growth, particularly in the larger

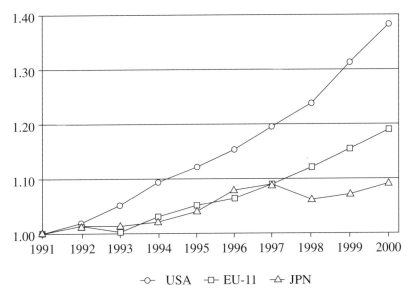

Figure 2.1 GDP 1991–2000 (1991 = 1)

European countries, to generate sufficient new job opportunities for unemployment to come down and new entrants to appear on the labour market. Indeed as Figure 2.2 shows, Europe witnessed a major employment crisis in the first half of the 1990s.

Combined, these two trends indicate that the substantial gap in aggregate economic welfare (real GDP per capita) between the US, the EU and Japan, some 30–20 per cent at the beginning of the 1990s has actually widened over the last decade, as illustrated in Figure 2.3. Again the GDP per capita levels in 1991 have been set at 1 so as to observe the trend between the EU relative to the US and Japan. Real GDP per capita in the US rose over the period 1991–9 by some 18 per cent, or 2.12 per cent annually. The EU countries saw real GDP per capita grow by some 13 per cent, or 1.62 per cent per annum, while Japan did not observe any rise in its real GDP per capita. From the EU countries, Ireland, Denmark and the Netherlands faced the largest increases in real GDP per capita over the 1990s.

The trends observed in Figure 2.3 are particularly striking when put in their historical context. To illustrate this, Figure 2.4 shows the relative growth performance, i.e. growth of per capita GDP, of a sample of OECD countries relative to the US. All four periods analysed, i.e. 1913–38, 1950–73, 1973–91 and 1991–2000, show a convergence pattern between the group of followers. The horizontal axis of Figure 2.4 measures the gap in

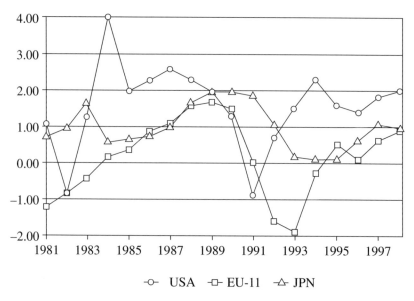

Figure 2.2 Employment 1981–1998

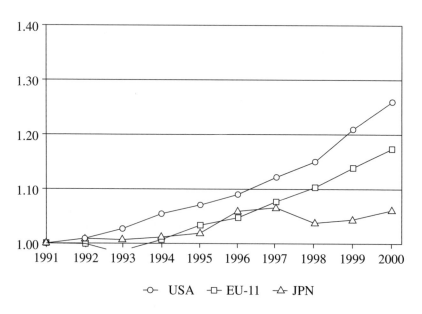

Figure 2.3 GDP per Capita 1991–2000 (1991 = 1)

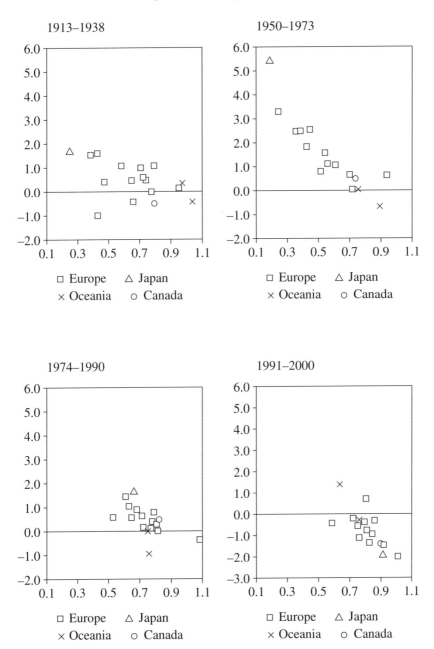

Figure 2.4 Convergence and divergence relative to the United States

the initial year. The vertical axis measures the growth rate of this gap for the relevant period. This gap is defined as

$$G_t = \frac{Y_{it}}{Y_{ust}}$$　　　　　　　(2.1)

which means that when the growth rate of G_t is positive, the gap with the US is falling, while a negative growth rate of G_t indicates that the gap with the US is rising. From Figure 2.4 it follows that in the first three periods the gap relative to the US is falling, i.e. countries are catching up with the US, while in the final period, 1991–2000, the gap with the US is increasing. For further analysis see Hollanders, Soete and ter Weel (1999).

Growth convergence *between* the various countries of the US, appears nevertheless a characteristic feature of the twentieth century. It reflects, through international trade, foreign investment, licenses and various other formal and informal information and knowledge channels, the continuous spreading of production, distribution and consumption patterns across the developed world. However, and as also shown in Figure 2.4, there has been a major shift from a general trend of catching up with the US to a new recent trend of the US suddenly increasing its lead over most other countries in the most recent period 1991–2000. As a parenthesis, it can be noted that this most recent period shows also convergence between the EU countries; at the same time the EU as a group has been falling behind the US.

From this perspective, it can be argued that the period up to 1973 was a period of rapid growth dominated by catching-up phenomena world wide: catching up of European consumption patterns to US standards; significant growth in the centrally planned economies based on further exploitation of 'Tayloristic' methods of labour organisation in agriculture and the heavy industrial sectors,[2] and the end of the de-colonisation process in most Third World countries. It was logical that through such a growth process the gap between the US and the EU and Japan would narrow down. By contrast, the US economy did, if anything, show some major weaknesses, for example in relation to employment creation.[3]

In contrast, the 1973–91 period appears to be characterised by the disappearance of such catching-up features, at least with respect to the European developed OECD world and Japan. This happened despite accelerated European economic integration with the subsequent enlargements of the European Community and the move from a customs union to an economic union. It also took place despite the gradual liberalisation of financial markets. In fact the period was characterised first by a dramatic explosion of exchange rate volatility, inflation, unemployment and public deficits. The failure of macroeconomic policies to contain inflation, reduce unemployment in Europe (with only one or two individual countries as exceptions) and control public spending is from this perspective both the consequence and cause of the halt in growth convergence and the end of what the French economic historian Forestier referred to as 'les trentes glorieuses': the thirty post-Second World War wonder years of high growth, low inflation and low unemployment.

Finally, the most recent period, from 1991 until 2000, appears as indicated in Figure 2.4d to be characterised by growth divergence between the US, Europe and Japan; effectively a leap forward by the US. This US growth divergence took place despite a major convergence between the US, Europe and Japan in aggregate economic indicators, such as inflation, long-term interest rates and public spending. It took also place despite the dramatic growth in international information and communication possibilities bringing about increased transparency.

It is hence important to re-situate in its recent historical context the continuing unexpected nature of this emerging growth divergence. First and foremost, few authors predicted the slow-down of Japanese growth. At the same time, many others predicted rapid growth in Europe because of the internal economic integration process deepening (the 1992 Single Market) and the expected rapid catching up of Eastern European countries to EU income and consumption levels. At the same time, the collapse of US growth predicted since the mid 1990s as a result of its low savings rate, high trade deficit and unsustainable growth in stock market prices failed to occur. One may thus conclude, and notwithstanding the recent growth upsurge in Europe, that for the first time in post-war history, growth divergence amongst the Triad countries has been a dominant feature of the 1990s.

Underlying the growth process over the last ten years, the question must hence be raised whether other, new factors appear to have emerged, particularly in the US. More than any other country in the world, the US economy appears to have benefited from faster application and implementation of new technologies, more rapid uptake of the new 'information highways' infrastructure and more successful world-wide commercial exploitation of these growth opportunities. In short, the US seems to have been the most successful country in making its transition to a knowledge-based economy.

2. THE EMERGING KNOWLEDGE-BASED ECONOMY

Before turning to particular trends in investment in knowledge in Europe, it seems essential to clarify somewhat the notions and concepts involved.

There is if anything today a conceptual confusion about what we have called here – using the older OECD term – the 'knowledge-based' economy, but what could be called as well the 'learning' economy (Lundvall and Johnson, 1994) or the 'new' economy. Knowledge is of course not a new concept. Its importance for economic growth was at the core of much economic thinking and writing of the late eighteenth and nineteenth centuries. One only has to think of the importance given to knowledge by classical economists, such as Marx or Schumpeter, to realise that economists, just as historians, have always been aware of the crucial importance of knowledge accumulation for long-term growth. What is new are a number of things. Without pretending to be exhaustive, we group them here under three headings: the economic integration in formal growth models of particular features of the knowledge accumulation process (as in new growth theory and beyond); the impact of new ICTs on the process of knowledge accumulation and hence also on growth ('new' economy and the emergence of open global electronic networks); and last but not least the increased importance of continuing knowledge improvements thanks to more routine use of a growing, enlarged base of codified knowledge (various interpretations of the 'learning' economy). Let us start with the first one.

Since the early 1980s, the economic profession has started to recognise the fact that knowledge accumulation can to a large extent be analysed like the accumulation of any other capital good. That one can apply economic principles to the 'production' and 'exchange' of knowledge; that it is intrinsically endogenous to the economic and social system and is not really an external, 'black box' factor, 'not to be opened except by scientists and engineers' as Chris Freeman (1972) once put it. Hence, while knowledge has some specific features of its own, it can be 'produced' and used in the production of other goods, even in the production of itself, like any other capital good. It also can be stored and will be subject to depreciation, when skills deteriorate or people no longer use particular knowledge and 'forget'. It might even become obsolete, when new knowledge supersedes and renders it worthless.[4]

But there are some fundamental differences with traditional material capital goods. First and foremost the production of knowledge will not take the form of a physical piece of equipment but generally be embedded in some specific 'blueprint' form (a patent, an artifact, a design, a software program, a manuscript, a composition) or in people and even in organisations. In each of these cases there will be so-called positive externalities; the knowledge embodied in such blueprints, people or organisations cannot be fully appropriated, it will with little cost to the knowledge creator flow away to others. Knowledge is from this perspective a 'non-rival' good. It can be shared by many people without diminishing in any way the amount available to any one of them. But of course there are costs in acquiring knowledge. A major central theme of economic theory is what is referred to as information asymmetry: the person wanting to buy something from someone who knows more about it obviously suffers from an asymmetry (a lack) of information.

It explains why markets for the exchange of knowledge are rare and why firms prefer in principle to carry out research and development (R&D) in-house rather than contracted out or licensed. It also provides a rationale for policies focusing on the importance of investment in knowledge accumulation. Such investments are likely to have high so-called 'social' rates of return, often much higher than the private rate of return. Investment in knowledge cannot be simply left to the market.

Second, the growing economic and policy consensus on the importance of knowledge for industrial competitiveness is undoubtedly also closely related to the emergence of the new ICTs. As already argued above, there is no reason to assume that the impact of new information and communication technologies would limit itself to the manufacturing and distribution of goods and services. Information technologies (ITs) are from this perspective in the real sense of the word 'information' technologies, the essence of which consists of the increased memorisation and storage, speed, manipulation and interpretation of data and information: in short what has been characterised as the 'codification' of information and knowledge. The additional Communication part in ICTs allows, however, such codified knowledge, data and information to become much more accessible than before to all sectors and agents in the economy linked to information networks or with the knowledge of how to access such networks. ICTs are hence likely to increase the rate of return to investment in knowledge, whether research, development, software or education expenditures. It is this particular feature which is identified with a possible higher 'new' growth path (see e.g., OECD, 2000; van Ark, 2000; Bartelsman and Hinloopen, 2000). Furthermore, such effects are likely to be most significant in countries where international access and transferability of codified knowledge has been difficult and costly.

It is from this perspective that the new ICTs have often been presented as the first case of truly 'global' technological transformation. The possibility of ICTs to codify information and knowledge over both distance and in real time brings about more global access. While the local capacities to use or have the competence to access such knowledge will vary widely, the access potential is there. ICTs in other words bring to the forefront the potential for catching-up, based upon cost advantages and economic transparency of (dis-)advantages, while stressing at the same time the crucial 'tacit' (that part of knowledge that cannot be codified) and other competence elements in the capacity to access international codified knowledge.

The importance of access brings to the forefront, on the one hand, the overriding importance of new communication infrastructures, not just for production and distribution but also for research and innovation, and, on the other hand, the crucial importance of the long-term availability of highly skilled manpower: not just scientists and engineers but more generally so-called 'knowledge workers': brainpower, which cannot be codified. Such human skills represent essential complementary assets to implement, maintain, adapt and use new physically embodied technologies. Human capital and technology are from this perspective two faces of the same coin, two non-separable aspects of knowledge accumulation.

Third, and as some authors in the field of innovation studies have argued, amongst them most explicitly Paul David and Dominique Foray (1995), as a result of the new ICTs the perception of the nature of the innovation process has also changed significantly over the last decade. Broadly speaking, innovation capability is today seen less in terms of the ability to discover new technological principles, than in terms of the ability to exploit systematically the effects produced by new combinations and use of aspects of the existing stock of knowledge, more widely and easily accessible than ever before. A similar set of arguments is made by Lundvall when he introduces his notion of the 'learning' economy. This model implies, as David and Foray have argued, to some extent more routine use of a technological base allowing innovation without the need for leaps in technology. It thus requires much more systematic access to the state-of-the-art. Universities, public and private research centres will introduce procedures for the dissemination of information regarding the stock of technologies available, so that individual innovators can draw upon the work of other innovators. The science and technology system is in other words shifting towards a more complex 'socially distributed' structure of knowledge production activities, involving now a much greater diversity of organisations having as their explicit goal the production of knowledge; what can be called learning entities. The old system, by contrast, was based on a simple dichotomy between deliberate learning and knowledge generation (R&D laboratories and universities) and activities of production and consumption where the motivation for acting was not to acquire new knowledge but rather to produce or use effective outputs. The collapse, or partial collapse, of this dichotomy conducts to a proliferation of new places having the explicit goal of producing knowledge and undertaking deliberate research activities. It raises obviously fundamental challenges as to the institutional adaptability of the still very fragmented and very 'dichotomised' European science and technology system. As argued in the fourth section this is precisely one of the areas where the US appears to have been more successful over the last decade.

It is important from this perspective to put the organisation of Europe's science and technology system in its own historical perspective. Science and technology has of course been the subject of national public interest and support for centuries. Large popular support for such activities was obtained by stressing the national security need and national prestige nature of such activities. At the European level, such national security or prestige arguments rarely developed, rather the contrary. Over the late 1970s and 1980s one witnessed a shift away from centralised public support for 'big science' areas considered of strategic importance, such as nuclear energy research (EURATOM) and aeronautics. The scope of such policies had been simple and in line with the early European integration aims: reap possible scale economies in production and large-scale research investment and secure European autonomy. Organisations such as CERN (European Council for Nuclear Research), ESA (European Space Agency), EMBL (European Molecular Biology Laboratory) became showpieces of European cooperation successes in science. With time passing though and the large nuclear energy and fusion programmes not developing their long-held economic promises, a new form of technology-based, industrial policies for new sunrise sectors, such as microelectronics, in the form of so-called 'pre-competitive' research support was developed. In their emphasis on competitiveness, such policies did of course overlap with national technology policies with the 'subsidiarity' principle more or less invented as belated legitimisation of existing flows of money and responsibilities between EU, national and in some member countries regional authorities. By adding some specific European networking requirements, such support policies did, however, respond to an apparent European need for more networking across different relatively closed research communities. Unfortunately, as such policies developed and started to eat up a growing and larger part of the European budget, the high-tech industries and large European firms which benefited most from such European pre-competitive research support programmes appeared to be those sectors and firms which came to lose most in terms of world market share. Hence, despite a growing European research budget, Europe has been losing ground most in international competitiveness in high-tech sectors.

Not surprisingly, the 1990s witnessed a significant shift in European policy making in the science and technology area. This shift can be best described as a shift in the nature of the public support: away from science and technology push towards more demand–pull policies, with greater acceptance of the crucial role of users and the intrinsic recognition that technical success does not necessarily imply economic success. The Commission's 'Green Book on Innovation' provided probably the most explicit recognition of the need for this shift towards innovation policies, describing Europe's failure in developing new products and new-technology based firms as a European technology paradox: excellence and strength in basic and fundamental research yet failure to translate this in commercial excellence and success.

Underlying this diagnosis and policy shift, there is, however, and in line with the argument set out above, a growing recognition that technical change in the highly developed, open societies of which Europe is composed is a complex dynamic process that involves many social and economic factors and a wide range of individuals, institutions and firms and hence is crucially dependent on effective networking between them.

From this perspective the capacity of Europe's economy to derive competitive advantage from technical change and innovation is more than ever dependent on the dynamic efficiency with which firms and institutions diffuse, adapt and apply information and knowledge. Within a context of fifteen different EU countries, each with their own, historically long-established science, technology and higher education institutions, long-term national networks of public and private research organisations and national policy makers in search of national prestige and recognition, this is a major policy challenge.

3.   IS EUROPE FALLING BEHIND IN KNOWLEDGE INVESTMENT?

From our previous discussion it will appear clear that there is no simple, agreed measure or notion, which could in a satisfactory way represent 'knowledge'. From what has been said above it will be clear that different components of capital and labour need to be taken into account in assessing the knowledge intensity of an economy. In a recent OECD report (1999), an attempt is made to bring together a number of knowledge intensity measures. Below, we compare some of these measures for the US, Europe and Japan,

In Figure 2.5, we compare the investment in tangibles, i.e. physical investment, and in knowledge as a percentage of GDP for the US, the EU and Japan for 1995. Japan has the highest physical investment intensity and the lowest knowledge investment intensity. By contrast, the US has the highest knowledge investment intensity and the lowest physical investment intensity. The position of the EU is in between. Countries such as Italy, Belgium, Germany and Austria have all knowledge investment intensity levels below the EU level; countries such as Sweden, France, Denmark, Finland and the UK have above EU knowledge investment intensity levels. If private spending on education and training had been included in Figure 2.5, the US figure would undoubtedly be even higher.

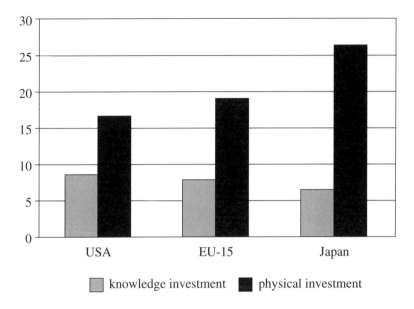

Figure 2.5 Knowledge and physical investment in 1995 (% of GDP)

Figure 2.6 represents the so-called ICT intensity – ICT expenditures as a percentage of GDP – for three different ICT components – IT hardware, IT services and software and telecommunications – for the US, Japan and the EU average for 1997. Of the individual EU member countries, only Sweden has now a higher level than the US and only Sweden and the United Kingdom have a higher level than Japan.

In Figures 2.7 and 2.8, the US efforts with respect to R&D expenditures are further compared with the expenditures for the EU in absolute, but real terms. This is done for the major two components where R&D is being carried out: so-called BERD, standing for Business Expenditures on R&D, in Figure 2.7 and GOVERD or government expenditures on R&D in Figure 2.8.

The trends in the Figures 2.8a and 2.8b are striking: an impressive gap between the US and Europe over the 1990s has emerged in privately carried out R&D. The US is today investing twice as much on business R&D as Europe; the current R&D gap is, as Figure 2.8b illustrates, more or less equivalent to the total EU R&D effort. By contrast in government carried out R&D the opposite is now true: Europe spends now nearly 25 per cent more on R&D than the US.

The data presented here without claiming to provide a complete picture are nevertheless striking in illustrating the way Europe appears to have

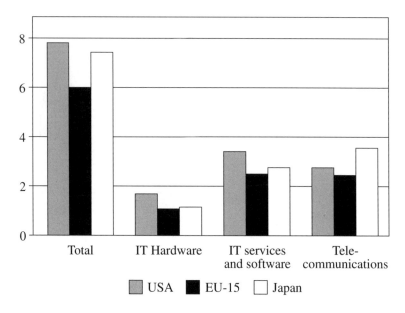

Figure 2.6 ICT expenditures in 1997 (% of GDP)

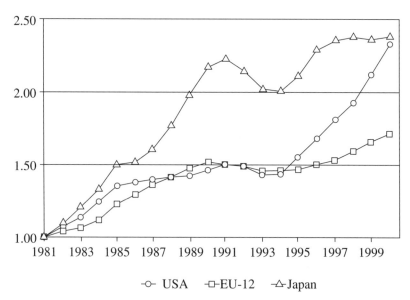

Figure 2.7a BERD 1981–2000 (1984 = 1) (Original series in 1990$PPP)

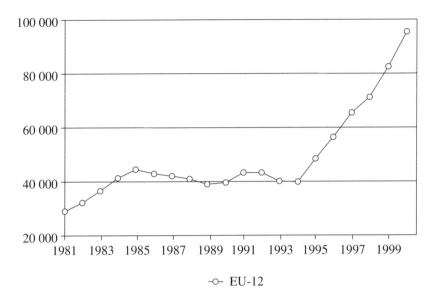

Figure 2.7b BERD US-EU-12 (1981–2000 in million 1990$PPP)

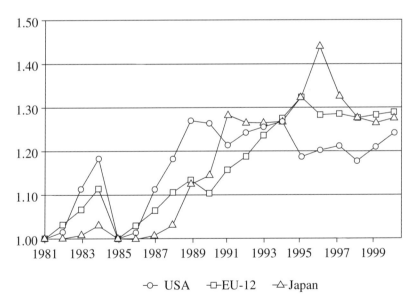

Figure 2.8a GOVERD 1981–2000 (1981 =1) (Original series in 1990$PPP)

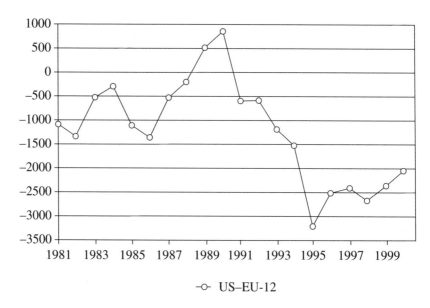

Figure 2.8b GOVERD US-EU-12 (1981–2000 in million 1990$PPP)

failed to invest in knowledge over the last decade: in R&D and ICT investment in particular. There is now a substantial gap in such investments in
the business sector between Europe and the US. On the face of the data presented here, Europe seems to have missed its transition to the knowledge-
based economy.

4. THE TRANSITION TOWARDS A KNOWLEDGE-BASED ECONOMY: WHAT CAN BE LEARNT FROM THE US EXPERIENCE?

In this last section we now turn in more detail to the US experience, as it
appears to have shifted in a more successful way its economic base in the
direction of a knowledge-based economy. In doing so we will follow the
three features that are at the core of the new importance given to knowledge accumulation as presented in the second section.

We start with the 'new' financial and economic recognition given to
knowledge as exemplified in new growth models. One may argue that in the
US in particular the early recognition by the financial world of the intrinsic value of knowledge accumulation has been behind a much greater readiness by the financial sector to invest in new, often purely knowledge-based

firms. From this perspective, the growth of an effective venture capital market whereby the resources to invest in knowledge accumulation could be extracted from the financial market appears to some extent to have been a crucial institutional innovation. Both in Japan and Europe (with the partial exception of the UK), the financial sector remained heavily 'material'-biased, often being formally involved or part even of the large industrial firms' management. Hence, the total market capitalisation of firms in the US has been much larger than in many continental European countries as compared for example, to their respective GDPs. Furthermore, the largest capitalised firms in the US are today practically all involved in knowledge accumulation activities rather than just material goods.

As Schumpeter in particular emphasised, the stock market was an essential institutional innovation accompanying the growth boom of the 1920s. It could well be argued that venture capitalists, the creation of NASDAQ and other financial innovations have allowed the mobilisation of private capital for investment in knowledge accumulation activities and have thus become an essential institutional innovation for the emerging knowledge-based economy. By the same token, at the macroeconomic level, given the importance of the stock market for additional income and earnings supplements (one may think of payments in options) for high-skilled knowledge workers, it could be argued that, consciously or not, a major macroeconomic innovation was introduced in the US by the Fed under the reign of Alan Greenspan. As a result, US monetary policy started to focus more on stock market developments than just money supply, hence signalling a shift in aim and purpose of the role of fiscal and monetary policy in a knowledge-based economy. Thus, while fiscal and budgetary policy is heavily restricted and focused on long-term policy aims (ageing of population, health costs, etc.), monetary policy has effectively been more relaxed and less related to inflationary fears, but rather enabling investments in new knowledge areas and further stock market valuation and expansion. The difference with what happened in Europe is from this latter perspective striking.

The second major feature in which the US appears to have been more capable of adjusting its economic structure to a knowledge-based economy is of course investment in ICTs. As illustrated in the previous section, investments in ICTs have been much higher in the US than in both Europe and Japan. Not surprisingly, the US appears to have been much more capable of benefiting from the emerging ICTs, contributing approximately to one-third of aggregate economic growth (US Department of Commerce, 1999), a figure, equal to the difference between US and EU economic growth over the 1990s as we saw in Section 1. There are several reasons each one insufficient in itself to explain the relative success of the US. Thus, there has been the dramatic revival of the US semiconductor industry following the US–Japanese semiconductor trade agreement, effectively providing breeding space for such revival. There has also been the successful alliance between software and semiconductor industry allowing for an effective commercial exploitation of technological improvements in the computer industry. Authors, such as John Zysman, refer in this context to the notion of 'wintelism': the combination of continuous technological improvements in chip performance (such as the Pentium from Intel) and in operating systems (such as Windows from Microsoft) requiring extensive performance capacity. Thanks to the combination of free local telecom access, expertise in hardware and software network technologies going back to DARPA and ATT (e.g., Sun and the software languages UNIX, Java and now Jini) and the development of a universal Internet Protocol (Netscape), Internet use rose rapidly outside of the traditional scientific community and was quickly taken up by businesses and individuals. Finally, the availability of extensive content (film, television, radio, press) provided a rapid take-off in terms of new Internet services.

The result has been that the US leads the world in Internet use and pricing, in number of websites, Information Service Providers, hits, sales on e-commerce, etc. The growth in employment in these ICT-related sectors has been significant, as has been the volume of international trade generated. The international US competitiveness in these sectors has undoubtedly been greatly enhanced by the imposition world wide of strong intellectual property regimes in the area of copyrights, trademarks and authorship rights.

But the final consumer end of these new ICT-based sectors is of course only one part of the 'new' growth story. Probably even more important has been the impact on firms' internal efficiency, the impact of so-called business-to-business e-commerce (OECD, 1998). The increased potential for codification and transferability allowed for by ICT enables also for significant reductions in transaction costs; for a process of des-intermediation and decentralisation of activities and more global direct distribution and access. Not surprisingly, the concept of the 'new economy' encapsulating some of these new growth features associated with ICTs has been primarily popularised in the US.

As third new knowledge feature, it could be argued that the US science and technology system has been much more successful in responding to the new, more complex knowledge production and distribution model associated with the new knowledge-based economy and described in Section 2.

First of all, as Nelson (1993) has described in great detail, the US national innovation system has traditionally been characterised by a university system, which was diverse (public–private; local–state; specialised–

broad), closely integrated with the private sector, and particularly strongly performing. Second, with the Bayh–Doyle act of 1981, the US has obtained strong first-mover advantages in valorising particular pieces of university research: one might say the 'commodification' of pieces of knowledge. Such a 'commodification' process has undoubtedly restricted the public use of university research results. Yet it has provided a major incentive to private firms to rely on university research for some of their own, too complex fundamental research activities and created a legally clear environment for university spin-offs, providing a major incentive for university staff to set up new private ventures. These trends are clearly reflected in the figures presented in Section 3 about Europe's private business knowledge investment gap. Europe has been lagging behind in allowing universities to take out patents (in most European countries, the individual university professors take out the patent) and shifting patent law to allow for university research to be patented. In fact, this has meant that in areas such as biotechnology and medical technology the leading position of some European countries has become eroded over the 1980s and 1990s.

As the figures in Section 3 illustrated, while awareness of both European and Japanese structural weaknesses in their innovation systems, and in particular their university systems, increased over this period, policy action continued to suffer from fragmentation and the dominance of the old national institutional bottlenecks. Thus, and particularly with respect to European Commission policy initiatives, the focus has continued to be primarily on fostering intra-European cooperation in the field of pre-competitive R&D, university researchers, and various support programmes for particular technology fields: the so-called framework programmes and other related technological support programmes. Unfortunately, compared to national resources the EU resources available were too limited to make any impact on shifting or redirecting countries' own national priorities in supporting investment in knowledge accumulation. At the same time, the policies seemed overly dominated by the overriding aim of intra-European research collaboration. While the latter is still welcome in specific cases, the essential research collaboration in the new complex knowledge production model is more likely to be of a global nature, going well beyond the European borders, and unlikely to allow itself to be described in terms of pre-competitive. Hence, there might even have been a case of knowledge acquisition 'diversion', the intra-European knowledge exchange and networking having taken place at the expense of extra-European exchange and networking.

Elsewhere (Soete, 1996), I called this a 'European paradox'. As Europe invested in intra-European research, in the collaboration and exchange of scientific knowledge among European scientists and in the technological strengthening of the competitive potential of European firms, the advantages of such geographically 'bounded' collaboration became marginal, given the dramatically increased opportunities for the fast international exchange of information and cooperation.

Obviously the analysis presented here in these few pages remains incomplete. Much more could be said about European performance in the three areas identified here. In some of these, such as Europe's higher education and university system, the picture drawn here, particularly in comparison with the US, is not so bleak. As the recent report of the US Council on Competitiveness (Porter and Stern, 1999) indicates, the US is also confronted with major policy challenges particularly in the field of federal support for long-term, basic R&D, weak secondary school performance and the declining pool of scientists and engineers engaged in R&D. The policy challenges raised by the transition to a knowledge-based economy are global and certainly not limited to Europe. Below we draw some policy conclusions for the European Union, which follow from our analysis.

5.   POLICY CONCLUSIONS

Three sets of policy conclusions emerge in a relatively straightforward manner from the analysis presented above. They can be organised along the three main 'new' features associated with the knowledge-based economy discussed in Section 2 and analysed in more detail with respect to the US in Section 4.

1.   The transition towards a knowledge-based economy requires a major investment effort in what can be defined today as knowledge investment: R&D, software, ICT hardware, telecommunications, education and training. While the amount of public support in Europe for some of these knowledge components – in particular education and government sponsored R&D – has been more or less in line with the US, the expenditures carried out in the business sector of the economy have been lagging behind dramatically over the 1990s. This holds for R&D, software, ICT investments and private spending on education and training. It is as if in Europe, both the business and financial community have not as yet discovered the economic value of knowledge investment. There are undoubtedly many reasons for this. Some might be culturally determined, others have to do with the performance of European financial markets. It appears likely that European financial markets are still suffering from their fragmentation. It is obvious that the European Union, particularly in areas such as venture capital, has

not yet realised the scale advantages of its large financial market. Yet the readiness to take risks, to provide capital to small and medium-sized economies (SMEs), to identify new investment opportunities in new areas are all closely related to the size of the market. All this calls for new priority setting in both the business and financial communities of Europe in knowledge investment. The fact that it is in particular private investment in knowledge activities which appears to lag behind what is happening in the US raises the question whether there is not a more systematic need for a new, common European fiscal regime for such expenditures. The US Council on Competitiveness makes a plea for making the R&D tax credit permanent so as to encourage higher R&D investment in the private sector. In many EU countries such tax credits do exist. However, they are often of a very different nature: in some countries limited to small firms, in other countries to particular regions. A common EU tax treatment of R&D would improve transparency and have an overall positive effect on R&D investment across Europe.

2. Europe's lagging *vis-à-vis* the US in the area of ICT investment and ICT use hides a much more diverse internal EU picture. Thus some member countries, such as Finland or Sweden, are world leaders in number of Internet hosts; others, such as Norway or Denmark, in number of household PC penetration; yet still others, such as Ireland, in ICT-related services growth, or in The Netherlands growth of mobile phones. Much can be learnt from these different European experiences, certainly when they are enlarged to include also the use of ICT in public and private services. There is a great potential here for cross-European learning. This includes systematic intra-European comparisons in the use of ICT in education (primary, secondary and higher), in health, in public administration, and in the many other public and semipublic areas of the economy, where competitive pressures are unlikely to bring about more rapid diffusion of ICT. There are clearly two sides to Europe's overall lagging behind the US in ICT. The first, more supply side set of arguments is linked to Europe's more general weakness in entrepreneurship and innovation. This has probably been even more the case in some of the most dynamic ICT areas: creating software tools, developing content industries. Europe's failure in these areas is hence not just the result of a lack of risk capital, of a cheap and transparent, common intellectual property regime, or of the dominance of national telecommunications monopolies, but of a broader systemic failure to develop new, fast-growing ICT and knowledge-based firms. The second, more demand side set of arguments emphasises the low adoption rate of ICT equipment and use. European

firms appear to have been much slower than their American and Japanese counterparts in learning how best to adapt their organisations to the new technologies. Yet again this aggregate picture hides a very diverse pattern. Comparing across Europe various forms and experiences of organisational learning is hence another area where significant insights can be obtained.

3. Both the European innovation and its science and technology system appear insufficiently adapted to the new challenges of the knowledge-based economy. There is here an urgent need for institutional adaptation. The EU can be instrumental in bringing about such institutional renewal, highlighting the overlapping and fragmentation of current national systems and inducing cross-European mobility. European networking in science and technology should aim at reaping both scale and scope advantages from existing European research activities. At the limit this does not require additional public research funds, but rather more efficient use, organisation and implementation of existing resources. The argument of a European research paradox hides to some extent the way, basic and fundamental research is still poorly organised in Europe, with national (sometimes regional) and European support programmes overlapping and widely different quality assessment of research performance. With European governments spending some 60 billion euro and the EU some 4 billion euro on R&D, national policy priorities continue to dominate the European research agenda and national institutions the implementation. There is, hence, a clear need for a new policy approach based on 'true' subsidiarity and networking aimed at reaping scale economies so as to improve European research quality. Compared to the US National Science Foundation (NSF) or National Institute of Health (NIH), Europe lacks so far the institutional framework to develop and implement its own research agenda. By contrast, at the innovation and technology level, apart from a number of specific regulatory areas, in which the setting of for example common technical standards and norms at the European level is likely to help firms in reaping scale economies, European technology policy should aim more at reaping some of the existing scope advantages of the large diversity and variety of current European research, as we already discussed above with respect to ICT use. Networking and subsidiarity are again essential to exploit such scope advantages, but rather of the opposite nature of the one described above. Thus the subsidiarity argument will rather put the emphasis on the contribution of local, regional authorities, and networking will aim at valorising cultural, historical and institutional differences so as to improve comparative learning and implementation. At the regional level, European

science and technology policies should try to foster the openness of local actors – universities, high schools, publicly funded research organisations, local SMEs, particularly in the high-tech sectors, and local subsidiaries of 'foreign' firms – whereas regional authorities' science and technology policies should aim at internalising the growth spill-overs of regional research. Both policy aims are clearly complementary.

We did not address here some of the other issues which emerged from our analysis and in particular the crucial importance of human capital investment for knowledge accumulation. This issue is more explicitly discussed in Robert Lindley's contribution to this book (1999).

POST SCRIPTUM ON THE LISBON EUROPEAN COUNCIL AND PORTUGUESE PRESIDENCY FROM THE PERSPECTIVE OF KNOWLEDGE INVESTMENT

The widely held view that the Lisbon summit was a success was largely based on the substantial agreements achieved on the two major Commission initiatives in the area of knowledge investment incentives as reported here: the proposals surrounding the so-called 'e-Europe' initiative putting forward with a clear time framework a number of specific ICT and Internet-use targets each member country should try to achieve; and second the concept of a European common research space proposing a policy line of action to overcome the many costs of 'non-Europe' in the area of research, development, innovation and entrepreneurship. Time will tell whether the broad policy agreement amongst the heads of states achieved at Lisbon on 23 and 24 March 2000 will be translated into effective policy action and ultimately lead to Europe becoming 'the most competitive and dynamic economy in the world'.

What is, with hindsight, particularly noticeable is that the Lisbon European Council took place at the height of the stock market's evaluation of Internet dot.com shares (the so-called technology, media and telecoms sector) and the US dominance of the 'new economy'. The poor growth and knowledge investment comparison of the EU with the US, also highlighted here, was hence not just extremely timely, it was also magnified. Since then, the dramatic decline in dot.com shares over the remainder of 2000, the growth slowdown in the US, the non-election of Al Gore – the Internet vice-president – have all contributed to a more sceptical view around the sustainability of the US new economy growth model. They have not, however, rendered the diagnosis of Europe's failure in knowledge investment any less instructive. Indeed, the slowdown of US growth and the likely slowdown of European investments in the US and of take-overs of US companies has so far not been accompanied by a major upsurge of such investments in Europe.

Nevertheless, what the recent US growth slowdown has brought to the forefront is that intra-European 'best practice' comparisons have become much more instructive particularly in the area of technology, innovation and information society policies, a line also defended here, and further endorsed in the 'Europe 2002 An Information Society for All Action Plan' prepared for the Feira European Council meeting (19 and 20 June 2000). The essential features of such benchmarking exercises are discussed at greater length in the contribution of Lundvall and Tomlinson. Suffice to emphasise here that benchmarking in this area underscores the particular strong growth and technology performance of some small 'new economies' in Europe and hence highlights also the possible, diverse routes towards the knowledge-based economy which Europe might take. For sure, and illustrated in the success stories of individual Nordic countries, there appears a Scandinavian knowledge-based economy or information society model based on trust between the citizen and the government and the efficacy of public services in embracing the new information and communication technologies. But there are other models, as the very different success stories of Ireland, the Netherlands and Portugal illustrate. Maybe it is in this simple fact: learning from Europe's diversity that one will find the secret of Europe's future success in the emerging knowledge-based economy.

ACKNOWLEDGEMENTS

This is the final draft of the Lisbon European Council follow-up book, edited by Maria João Rodrigues. I am particularly grateful to Bas ter Weel for help with the tables and figures and to Robert Boyer, Gøsta Esping-Andersen, Robert Lindley and Maria João Rodrigues for critical comments on previous drafts.

NOTES

1. The first draft of this paper was written in May–August 1999 in preparation of the Lisbon summit (23 and 24 March 2000), a number of European policy initiatives have been taken since. That is why we add a short policy Post Scriptum on the outcome of the Portuguese Presidency.
2. The impulse to growth under communism would become based on the electricity revolution and the scientific Tayloristic division of labour organisation.
3. See e.g., the various contributions to the so-called Automation debate (US Senate Committee, 1966).

4. There are today an innumerable number of studies which have articulated in quite some detail the economic principles of knowledge and knowledge accumulation. For overviews see amongst others Lundvall and Borras (1997) and the contributions edited by Archibugi and Lundvall (2001).

REFERENCES

Archibugi, D. and B.A. Lundvall (eds) (2001), *The Globalising Learning Economy: Major Socio-economic Trends and European Innovation Policy*, Oxford: Oxford University Press.

Ark, B. van (2000), *Measuring Productivity in the 'New Economy': Towards a European Perspective*, The Economist (148) 1, 87–105.

Bartelsman, E. J. and Hinloopen, J. (2000), *ICT en economische groei: een hypothese*, Economisch Statistische Berichten (85), 4254, 376–378.

Borrus, M. and J. Zysman (1997), 'The rise of Wintelism as the future of industrial competition', *Industry and Innovation*, pp. 141–66 (2, December).

David P.A. and D. Foray (1995), 'Accessing and expanding the science and technology knowledge base', STI outlook, *STI Review*, 16, Paris: OECD, pp. 21–2.

Freeman, C. (1972), *The Economics of Industrial Innovation*, Penguin.

Hollanders, H., L. Soete and B. ter Weel (1999), 'Trends in growth convergence and divergence and changes in technological access and capabilities', MERIT Research Memorandum 99-019, *http://meritbbs.unimaas.nl/rmpdf/1999/rm99_019.pdf*

Robert Lindley's contribution to this book.

Lundvall, B.A. and S. Borras (1997), *The Globalising Learning Economy: Implications for Innovation Policy*, European Commission, DG XII, Brussels.

Muldur, U. (1999), *L'allocation des capitaux dans le processus global d'innovation est-elle optimale en Europe?*, European Commission, DG XII, Brussels.

Nelson, R.R (ed.) (1993), *National Systems of Innovation: A Comparative Study*, Oxford: Oxford University Press.

OECD, *Science, Technology and Industry Scoreboard 1999: Benchmarking Knowledge-based Economics*, OECD: Paris, OECD.

OECD (1998), *The Competitive Dynamics of Internet-Based Electronic Commerce*, Paris.

OECD (2000), *A New Economy? The Changing Role of Innovation and Information Technology in Growth*, Paris, OECD.

Porter, M. and S. Stern (1999), *The New Challenge to America's Prosperity: Findings from the Innovation Index*, Washington DC: Council on Competitiveness.

Soete, L. (1996), 'New technologies and measuring the real economy: the challenges ahead', paper prepared for the ISTAT Conference on 'Economic and Social Challenges in the 21st Century', Bologna, 5–7 February.

Soete, L. (2000), *Toward the digital economy; Scenarios for business*, Telematics and Informatics 17 199–219.

US National Commission on Technology, Automation and Economic Progress (1966), 'Technology and the American Economy', Report to the US Congress, 6 Volumes, Washington.

US National Commission (1966), *US National Commission on Technology, Automation and Economic Progress*, Washington DC.

3.   A new European social model for the twenty-first century?

Gøsta Esping-Andersen

The history of European welfare states has combined rare moments of epochal change with long spells of politics as usual. It is now more than one hundred years ago that Bismarck launched modern social insurance, and a half century has elapsed since today's welfare states were carved out of war-torn Europe. Both instances stand out because the architecture of the state was fundamentally recast, because visionary thinkers and bold statesmanship embraced new ideals of social justice. But, for the most part, social policy has simply meant incremental fine-tuning and adaptation to the existing edifice. In normal times, social policy is mainly conducted by bureaucrats and technicians.

Epochal redefinitions belong to periods when our basic goals must be reconsidered. Gustav Moeller's and Lord Beveridge's designs for a modern welfare state were brought forth by the urgency of consolidating democracy and new social solidarities. While democracy is now an unquestioned reality in Europe, social cohesion is not. We are moving towards a new economy and society, both of which call for a new social model. Europe, today, stands at a cross-roads similar to the era when we invented the postwar welfare state. If the burning issue is to better align redistributive priorities and social rights to the evolving reality, this is not the right moment for bureaucrats or technicians to reign. What I here present is therefore meant as an effort to rethink the *Gestalt* of the European welfare state; an attempt to construct a welfare edifice that is in better harmony with the kind of economy, employment and family that is in the making.

1.   THE CHALLENGE: AN OVERVIEW OF THE ISSUES

The challenge is immense because the revolutions in both employment and family structure create fantastic new opportunities but also new social risks and needs. Changing technologies, intensified global integration, and the need to upgrade our human capital are paramount in order to ensure competitiveness. But the emerging economy will not be solely knowledge intensive. Services will dominate employment, and this means by and large a bias in favour of professional and technical occupations and skills. Yet, in response to families' new needs, we will also see the growth of lower-end (and potentially low-paid) personal and social services that, furthermore, are vital for any realistic full-employment scenario.

It will, accordingly, be difficult to avoid new dualisms. A knowledge-intensive economy will produce new skill-based cleavages. How to deal with the potential losers is one major and urgent challenge. A knowledge-intensive economy necessitates not just expertise among producers, but also among consumers. Therefore, unless Europe succeeds in broadly strengthening the cognitive capacities of its citizens, one ominous, long-term scenario is a smattering of 'knowledge islands' in a great sea of marginalised outsiders. Europe's challenge is not simply to catch up with the United States' lead in information technologies, but also to ensure that this will not endanger social cohesion. This poses the first-order challenge of how to democratise skills through education and training.[1] But, a society's ability to develop its human capital resources to the full depends also on social policy.

The more simple-minded 'Third Way' promoters believe that the population, via education, can be adapted to the market economy and that the social problem will, hence, disappear. This is a dangerous fallacy.[2] Education, training or life-long learning cannot be enough. On one hand, a skill-intensive economy will breed new inequalities. On the other hand, investments in education can be inefficient if they are not backed up by social investments: children's ability to learn and succeed in school depends directly and powerfully on the social situation within their families. Lingering social inequalities invariably produce educational and cognitive inequalities. And the more we progress towards the knowledge economy, the higher will be the social cost of such inequalities.

A second formidable challenge comes from new family and life-course patterns. Families are increasingly unstable, and new household forms are emerging. In their wake come new risks, new needs. Yet again, we see signs of growing polarisation between the strong and the weak. At one end, divorce, separations and single parenthood create risks of poverty. At the other end, dual-earner households strengthen families' resource base. Since marital homogamy is the norm, inequalities between family types will rise.

The standard male breadwinner model that once guaranteed adequate welfare and high fertility is declining both numerically and in its capacity to effectively prevent child poverty. Indeed, the conventional family may increasingly constitute an obstacle to flexibility and adaptation since too many citizens' welfare depends on the job and income security of one person. And with family revolution is emerging new life-course patterns, much less linear, homogeneous and predictable. As a result, new risks are bundling heavily among youth and in child families. The challenge, again, is to redefine social policy so that it nurtures strong and viable families and protects those most at risk. If those most at risk happen to be children and youth, the urgency of reform is so much greater because it is today's children that will be tomorrow's productive base – or, in the case of failure to reform, tomorrow's expensive social problems. For a knowledge-intensive society, there is one clear social policy guideline that emerges: *give absolute priority to assuring the welfare of children.*

As was always the case, access to paid work is families' single best welfare guarantee. Emerging household patterns all mean that the family–work nexus has become more problematic. We see the growth of 'no-work households' (about 10–12 per cent of all working-age households in the EU now), of young adults continuously depending on their parents, and of single parent families. We also see a massive rise in two-income, two-career households. In each and every case, the bottom line is that households encounter, often severe, incompatibilities between employment and family obligations. Such trade-offs cannot easily be resolved within the family itself, nor within the market. The modern family requires access to services in order to avoid dependency on traditional income maintenance. The household and the family are the bedrock of social cohesion, but in their modern form they face severe dilemmas. And women's employment is emerging as a *sine qua non* for both welfare maximisation and as a vital economic resource. A second clear policy guideline therefore emerges: *prioritise services to households in order to reconcile mothers' career and family objectives.*

Post-war welfare capitalism functioned well because labour markets and families themselves were the *principal source of welfare* for most citizens, most of their lives. At present, both labour markets and families create widespread insecurity, precarity. The consequence is easily social exclusion. The social policy instruments that are needed to combat social exclusion in the twenty-first century are, in large measure, the very same that are needed to secure a cognitively strong workforce: investments in families and children.[3]

The highly aged and income transfer-biased emphasis of most contemporary European welfare states is not an efficient architecture for dealing with the new profile of risks and needs. A re-design of social policy implies a re-thinking of the life cycle, of the balance between income transfers and services and, more generally, of the guiding principles of social justice and equality. If Europe aims to strengthen its competitive position in the world economy and, at the same time, commit itself to full employment, new inequalities will be difficult to avoid. The burning question is, what do we do about them? The most fundamental conclusion that emerges is that we must re-think the concept of social rights. The existing principle of guaranteeing maximum welfare and equality for-all-here-and-now cannot be consistent with emerging economic imperatives. If relatively low incomes, bad jobs or precarious employment cannot be avoided (and might arguably even merit encouragement), there is the issue of how to soften the welfare effects in the short run. However, the core welfare issue must focus on the dynamics of peoples' lives, on citizens' life chances. Low wages or bad jobs are not a threat to individuals' welfare if the experience is temporary; they are if individuals become trapped. In brief, the essence of social rights should be re-thought as effective guarantees against entrapment, as the right to a 'second chance'; in short, as a basic set of *life chance guarantees*.

2. THE DIVERSITY OF EUROPEAN WELFARE REGIMES

These challenges are not equally severe across all European welfare systems. We must avoid two errors. One is to ignore the great diversity of European welfare systems. A second is to remain too narrowly preoccupied with just the welfare *state*. Society's total welfare package combines inputs from the welfare state proper, markets (and especially labour markets) and families. Many view the welfare state as overburdened, inefficient, threatened or, simply, malfunctioning. Some advocate that it be radically slimmed, others that it be strengthened and still others that it be overhauled. Whatever opinions are put forth, there is an implicit view of what, alternatively, ought to be the role of markets and families. Those who advocate 'decentralisation' basically suggest a greater responsibility to families and the 'local community'; those who champion privatisation assign welfare to the cash nexus but, in practice, the result would also be a greater burden on many families. To capture the interplay of state, family and market, it is useful to cast our analysis in terms of welfare *regimes*.[4]

Turn the clock back to the post-war decades, and we would identify two distinct European welfare regimes. The Nordic-cum-British was largely general revenue financed, stressing universal, flat-rate benefits. The other, prevalent in Continental Europe, emphasised contribution-financed and employment-based social insurance. As social protection systems evolved and matured by the 1970s, differences emerged much more clearly. The Nordic countries branched out into a unique model, first by adding an earnings-related component to flat-rate 'citizens' benefits and, secondly, by shifting the emphasis from cash transfers to servicing families, stressing employment-activating policies and, above all, women's integration into the labour market. The Nordic model may be famous for its generosity and universalism, but what really stands out is its employment bias and its 'de-familialisation' of welfare responsibilities. Britain, in contrast, gradually moved towards more income testing, and also more market reliance – thus converging with North America. The hallmark of most Continental European countries is how little has changed. They remain firmly wedded to employment-based, contributory social insurance but have extended coverage to residual groups via *ad hoc* income-tested programs (like the RMI in France or the pensione sociale in Italy). A second defining feature of Continental European, and especially Mediterranean, social protection systems is their strong *familialism*, i.e. the idea that families hold the principal welfare responsibility for their members, be it in terms of sharing incomes or caring burdens. The upshot is a system which is committed to protecting the male breadwinner (via insurance and job protection), highly reliant on social contributions for financing, and comparatively very underdeveloped in social services.

Such differences mean that we cannot forge general strategies for social reform at an abstract pan-European level. It also follows that we shall err terribly if we limit our attention solely to *governments'* welfare role. I believe it is futile to discuss whether we should reduce (or increase) public responsibilities without considering their interaction with family and market welfare delivery. A strategy of 'de-centralising' welfare may sound appealing to many, but how will it affect women's double role as workers and caregivers? Alternatively, a scenario of more markets may appear more efficient, but if large populations will be priced out of the welfare market, do potential efficiency gains clearly outweigh potential welfare losses? Reforming European welfare commitments for the coming century implies *regime change*, that is reordering the welfare contributions of markets, families and state so that the mix corresponds better to the overall goals we may have for a more equitable and efficient social system.

3. THE TRANSFORMATION OF THE SOCIAL RISK AND NEEDS STRUCTURE

Most European social protection systems were constructed in an era with a very different distribution and intensity of risks and needs than exists today. With the main exception of Scandinavia (and Britain), the allocation of welfare responsibilities between state, market and families has not changed dramatically over the past fifty years. What has changed, however, is the capacity of households and labour markets to furnish those basic welfare guarantees that once were assumed to be their domain. Indeed, both now generate new risks and, equally importantly, also new needs.

The post-war model could rely on strong families and well-performing labour markets to furnish the lion's share of welfare for most people, most of their lives. Until the 1970s, the norm was stable, male breadwinner-based families. With few interruptions, the male could count on secure employment, steady real earnings growth, and a long career – followed by few years in retirement after age sixty-five. Women would typically cease to work at first birth, and were thus the main societal provider of social care. As such, the standard, prime-age household could be assured of adequate welfare. The problematic risks of this epoch bundled at the two 'passive' tail-ends of the life cycle: in large child families and among the aged. Not surprisingly, post-war welfare states came to prioritise income maintenance and, *par excellence*, pensions.

If, as now, new risks emerge from weakened families and poorly functioning labour markets we confront a concomitant market and family 'failure'. And the consequence is that the welfare state becomes burdened with responsibilities for which it was not designed. A well-functioning welfare state for the future must, accordingly, be re-calibrated so that labour markets and families function more optimally.

3.1. Family risks

A paradox of contemporary society is that child poverty is rising even when families now have very few children. Ongoing changes affect young households most severely. The reasons are well documented. Firstly, unemployment and precarity is concentrated among youth and the low educated (males in particular). Homogamy means that unemployment, precarity and poverty 'comes in couples'. Youth often face serious delays in 'getting started', in making a smooth transition from school to careers, or in forming independent families; Southern European youth can often anticipate three years' unemployment and this, obviously, is one cause of falling birth rates. Nonetheless, the consequences of youth precariousness vary depending on national social policies. As seen in Table 3.1, the unemployed – particularly youth – face severe revenue problems in many EU countries. Southern Europe's 'familialism' implies that the unemployment problem is internalised within the family of origin, but this is not the case in Northern Europe. Where, as in Denmark, unemployed youth are typically entitled to social benefits, poverty is modest; where, as elsewhere, they rely primarily on assistance, poverty is widespread.[5]

The new risks are also related to the rise of 'non-standard' households. For an overview, see Table 3.2. Two types have, in particular, become

Table 3.1   Per cent unemployed individuals with zero market or welfare state revenue, and dependence on familial support

	% with zero income ALL (20–64)	% with zero income Youth (20–30)	% with income from family Youth (20–30)	% living with parents Youth (20–30)
Belgium	2.1	3.3	13.7	28.8
Denmark	0.4	0.0	12.6	7.8
France	13.7	20.8	4.8	42.3
Germany	5.4	8.5	15.5	31.0
Greece	48.9	59.6	7.8	55.1
Ireland	2.1	3.0	0.4	67.8
Italy	69.8	81.3	4.2	82.9
Portugal	40.7	54.2	0.0	57.9
Spain	31.8	42.6	2.6	66.3
UK	3.6	2.2	5.2	39.7

Source: Own analyses of 1994 Europanel. See Bison and Esping-Andersen (2000).

Table 3.2   Per cent non-standard working-age families and their economic situation (mid-1990s)

	No work-income households % of all	Relative disposable income of no-worker households (all = 100)	Single parent households % of all child families	Poverty rate in single-parent households[a]
Belgium	13	64	20	13
Denmark	8	63	18	16
Finland	7	60	12	5
France	12	67	16	23
Germany	12	57	16	41
Ireland			18	
Italy	10	51	15	48
Netherlands	14	61	16	33
Portugal			13	
Sweden	8	59	18	7
Spain			13	
UK	13	56	21	49
USA	6	39	27	54

Note: [a] Poverty is measured as <50 per cent median, adjusted income (equivalence scale = 0.5).

Source: OECD.

Table 3.3 *Poverty rates in child families after taxes and transfers (mid 1990s)[a]*

	Two adults, one earner	Two adults, two earners	Lone parent, not working	Lone parent, working
Belgium	2.8	0.6	27.6	11.4
Denmark	3.6	1.0	17.7	10.0
Finland	3.5	1.5	37.1	3.0
France			45.1	13.3
Germany	5.6	1.3	49.5	32.5
Greece	15.1	5.0	36.8	16.3
Italy	21.2	6.1	49.1	24.9
Netherlands	4.7	1.2	22.6	17.0
Sweden	6.0	0.8	24.2	3.8
UK			69.4	26.3
USA	30.5	7.3	67.0	38.6

Note: [a] poverty = <50% of median income, using equivalence elasticity = 0.5.

Source: OECD.

prominent: the 'no work income' and the single-parent household.[6] Both run high risks of income poverty. No-work households are generally transfer dependent, often relying on social assistance. Except in Scandinavia, child poverty is alarmingly high in lone-parent families. Yet, across *all* kinds of child families – in two-adult families as well as in single-parent households – the best safeguard against poverty is not so much transfer generosity, but that mothers work (see Table 3.3). The low levels of single-parenthood poverty in the Nordic countries are, in fact, due less to generous social transfers than to adequate work incomes made possible by child care.[7] Simply put, mothers' employment is a very effective antidote to the risks that come with family instability and labour market precarity. If this is the case, the really pressing social policy issue has less to do with income maintenance and much more to do with servicing working mothers.

The new distribution of life-cycle risks is most evident when we contrast younger and older households. Tables 3.4 and 3.5 show that the economic well-being of child families has been eroding while, concomitantly, it has improved among the elderly. High incomes have allowed the elderly to live independently and, coupled to rising longevity, this implies that the chief needs among the aged are shifting towards care services in advanced age. And herein lies one of the key epochal transformations: the main welfare needs within young *and* aged households have less to do with improved

Table 3.4   Trends in relative disposable income by household type
             (Percentage point change, late 1970s–mid 1990s)

	All households with children	Young households	Retired households 65–75
Austria	−3	−1	+11
Denmark	−3	−8	+5
Finland	+3	−9	+1
France	0	−5	0
Germany (W)	−3	−2	−3
Italy	−1	−3	+3
Netherlands	0	−7	−4
Sweden	−2	−11	+9
UK	−4	−3	+5
USA	−3	−5	0

Source: OECD.

Table 3.5   Trends in after-tax/transfer child poverty rates and in the ratio
             of child-aged poverty incidence, 1980ca.–early 1990s

	Child poverty 1980	Child poverty mid-1990s	1980 child-aged poverty ratio	1990s child-aged poverty ratio
Belgium		5		0.45
Denmark		4		0.67
Finland		3		0.21
France	13	17	0.81	1.60
Germany	5	11	0.45	1.10
Netherlands	7	9	1.17	1.29
Portugal[a]	28	23	0.67	0.53
Spain[a]	17	17	0.53	0.68
Sweden	2	3	0.33	1.00
UK	11	27	1.22	2.25
USA	23	30	1.05	1.50

Notes: [a] last year = 1988.
         Poverty = <50% of median adjusted household income after taxes and transfers
         (equivalence scale = 0.5).

Source: LIS data.

Table 3.6   The servicing and age bias of welfare states, 1992

	Service bias: ratio of Services/cash	Aged bias: ratio of youth/aged
Austria	0.03	0.31
Belgium	0.08	0.61
Denmark	0.33	1.37
Finland	0.21	0.94
France	0.13	0.48
Germany (W)	0.12	0.43
Ireland	0.16	1.12
Italy	0.07	0.79
Netherlands	0.12	0.79
Portugal	0.10	0.55
Spain	0.05	0.67
Sweden	0.45	0.92
UK	0.13	0.50
USA	0.06	0.30

Note: Health care is excluded from services.

Source: OECD Social Expenditure Database.

income transfers and more to do with access to services. Among the ultra-aged in particular, the pressing need is for home help services and residential care. Within child families, poverty is best stemmed by enhancing parents' labour market prospects and earnings capacity.

Yet, most European welfare states remain uniquely biased in favour of the aged rather than youth, in favour of income maintenance rather than services. Herein lies a growing asymmetry between private need and public provision. The Nordic countries are unique in their attempt to prioritise services to young families (see Table 3.6).[8] Put differently, most European nations may be overspending on passive maintenance and underinvesting in the kinds of resources that strengthen citizens' capacities to be self-reliant.

Services can, of course, be provided also by the market or by families themselves. In Europe, however, marketed family services, such as private day care, are generally priced out of the market.[9] In brief, where government provision (or subsidisation) is absent, as in most Continental and especially Southern Europe countries families themselves must shoulder most of the caring burden of children and the elderly. The upshot of familialism is of course to worsen women's inability to harmonise family responsibilities and paid work and, indirectly, to weaken families' capacity to autonomously combat poverty. But since, in effect, today's young women do work, traditional familialism mainly provokes another perverse result, namely fewer children than families actually desire. With fertility rates between 1.2 and 1.5, most of Europe finds itself in a *de facto* low fertility equilibrium. Considering that the average European desires a little more than two children, such fertility levels are a clear signal of widespread dis-welfare.

4. WELFARE ASYMMETRIES ACROSS GENERATIONS AND THE LIFE COURSE

There is a rising tendency to view the welfare of the aged and the young as a basic trade-off. This can be dangerous. There exists, of course, some evidence that the welfare improvements of retirees occurs at the expense of youth and children, at least in countries (like the US or Italy) where more generosity to retirees has not been accompanied by an upgrading of family policies (Preston, 1984; Easterlin, 1987; Palmer, Smeeding and Torrey, 1988; OECD, 1998; Mirowsky and Ross, 1999). Also, it is clear that income distribution trends in most countries favour the aged. As shown in Table 3.7, the median retired household can usually count on a disposable income of at least 80 per cent of the national median.[10]

Table 3.7   Level of relative disposable income of persons aged 65+, 1990s (per cent of median)

	Disposable income
Belgium	77
Denmark	73
Finland	78
France	94
Germany	86
Italy	84
Netherlands	85
Sweden	88
USA	92

Note: Disposable income is net of taxation and transfers, and includes all income sources.

Source: OECD Income Database.

Certainly there remain pockets of poverty among the aged, typically concentrated among widows and persons with problematic contribution histories. Old-age poverty tends to be higher in countries with recently large rural populations (such as Greece, Italy and Spain). It is also well known that retirement income declines somewhat with age. Nonetheless, all indications are that the large mass of pensioners in most countries have sufficient (and sometimes perhaps 'excess') incomes, especially in light of reduced consumption and household capital expenditures, and because an often very large proportion (the EU average is 75 per cent) of the elderly own their home outright. What is more, in many countries retirees enjoy preferential tax treatment and are generally exempt from social contributions.

The economic well-being of today's elderly is the result of a unique combination of factors that produced high retirement income and life-time asset accumulation.[11] OECD (1998) shows that the average household at age sixty-five possesses wealth that equals four to five times its annual income stream. And, although we have only scattered nation-specific evidence, there are indications of pension overprovision in some countries. My own analyses of Italian family expenditure data indicate a 30 percentage point excess of income over expenditure in the average retiree household. A recent study by Kohli (1998) on intra-family money streams indicates a huge dominance of transfers from the aged (seventy plus) to their children and grandchildren: 24 per cent of income is transferred to their children; almost 15 per cent to their grandchildren.[12]

Such downward intra-family redistribution surely varies by income decile and by nation. Moreover, excess revenues reflect not just pension generosity but also home ownership, private assets and lower consumption needs. Still, the redistributive effect must be considered perverse if the welfare of youth is becoming a function of the riches of their retired forebears. Indeed, it is doubly perverse in the sense that pay-as-you-go pensions are financed by the working-age population. The welfare state was presumably built in order to even the playing field, but here is a case where it helps re-establish inherited privilege.

Any debate on reforming pensions must consider the life-course specificities of past, current and *especially* future retiring cohorts. If current retirement cohorts are generally well off it is because they are the main beneficiaries of Golden Age capitalism. Firstly, most of their careers spanned decades of strong productivity and earnings growth, with low unemployment among prime-age males. Secondly, with the regulation of seniority rights and the emergence of efficiency wage systems, the age–wage profile was de-coupled from productivity – hence rising earnings even when productivity declines with age.[13] Thirdly, today's pensioners are the chief beneficiaries of pension upgrading in the 1960s and 1970s. Fourthly, although there has been a decline in real earnings growth in recent decades, the financial returns on investments have risen.

A major reform of present pension systems confronts the dilemma that future retirement cohorts may not amass similar life-time assets, either through individual initiative (work and savings) or through the redistributive mechanisms of public pension schemes. Or, more likely, future retirement cohorts will, if uncorrected, become far more dualistic, possibly even polarised in terms of life chances. Today's youth often face serious delays in passing to stable employment: besides longer schooling, a large proportion can anticipate protracted and maybe frequent unemployment combined with more precarious employment. This correlates with skills and education.[14] Secondly, as de-regulation weakens the security of the prime-age 'insider' work force, career interruptions and redundancies are increasingly likely, among the less skilled in particular.[15] Thirdly, today's young cohorts are unlikely to benefit from decades of powerful real earnings growth and, if productivity bargaining becomes increasingly decentralised, seniority-based wage systems may weaken. Again, there is a clear trend towards more inequality in skills-based earnings power. Fourthly, those who are young today will fully experience the impact of ongoing pension reform in EU member countries, with the shift towards more individualised and actuarially based entitlement calculations. And, yet once again, this will favour the strongest workers in the labour market.

If *de facto* retirement age will remain at fifty-nine to sixty, today's young cohorts will be hard put to cumulate a minimum of, say, thirty-five contribution years towards the basic pension. These cohort-specific disadvantages are, nonetheless, offset by three key factors: first, their higher educational attainment and superior cognitive skills imply greater adaptability and ability for retraining across their careers. As they age, an investment in retraining may appear more logical than, simply, early retirement. The stronger the skills and educational base of young workers today, the greater will be the pay-off when they eventually age.[16] Secondly, each new retirement cohort shows sharp improvements in health and longevity, and all indications are that this will continue. Already today, the typical sixty-five-year old male can expect another eight to ten disability-free years (OECD, 1998). Those who are young now will be able to count on many more disability-free years. Thirdly, the ongoing growth of women's life-time employment implies that future retirement households will be able to double up pension savings or, in the case of divorce, women will increasingly have independent pension entitlements.

A strategy of resolving the looming pension crisis by radically reducing pension entitlements today would be counter-productive in the long run if, as is very possible, future retirees will look more like their forebears in the 1940s or 1960s. If, now, pensioner households have too much income, it would be a more equitable, and certainly more prudent, policy to simply tax away the excess.[17] If, then, a major reduction of public pensions is a sub-optimal long-run strategy, our attention must shift to alternative policy. As virtually all agree, the key to long-term sustainability lies in population growth and, more realistically, in raising participation rates (OECD, 1996; 1998; Thompson, 1998; Orszag and Stiglitz, 1999).[18]

4.1. Fertility and immigration

Population ageing is simply the effect of fertility decline and longer life expectancy. The great paradox of contemporary Europe is that birth rates are positively correlated with women's employment. Many EU nations, and in particular Southern Europe, display the characteristics of a long-run, *low fertility equilibrium*. Certainly, it will make some difference for the long run if fertility is, like in Denmark, at 1.8 or, as in Italy and Spain, at 1.2. But a return to fertility rates at 2.0 or more will not make any appreciable difference in terms of the ratio of actives to aged by 2030 or 2050 (OECD, 1998).[19] The same conclusion emerges from assessments of the effect of predicted, realistic immigration rates: resolving the retirement burden through population growth will not make a huge difference for the coming fifty years – unless, perhaps, immigration expands markedly over several decades (OECD, 1998). To give an idea of the magnitudes required, a recent simulation study concludes that (in the US case) the present annual immigration rate would have to double (or, if a family-unification policy were pursued, quadruple) *and* immigration would have to be limited to educated males in the forty to forty-five age range, in order to ensure long-term financial equilibrium (Storesletter, 2000).

Nonetheless, a permanent low-fertility equilibrium will worsen age dependency ratios.[20] It will have serious consequences for macroeconomic growth and, perhaps most importantly, it is a symptom of welfare problems within contemporary families. The simple fact is that families are no longer capable of, or willing to, internalise the full costs of children. Therefore, if children are a priority welfare goal in their own right, EU member states cannot avoid addressing the dilemma of how to allocate the costs of children. The socialisation of a large part of the costs in Scandinavia is undoubtedly a major reason for why families there approach actually desired fertility levels. It is vital to recognise that a knowledge-intensive society must maximise its future human potential in order to compete. It follows that the costs of children are not solely current consumption outlays but also a capital investment with, possibly, huge future returns –

some of which benefit the individual and some the collectivity. Accordingly, as I shall argue below, a central component of any workable win–win policy must be to prioritise investment in children and youth, to redistribute the costs of children and to mobilise better our actual human capital reserves.

4.2.   Participation and employment

European employment rates diverge dramatically. Female rates range from 70+ per cent in Scandinavia to less than 40 per cent in Ireland, Italy and Spain. The difference declines noticeably among younger cohorts, but these also suffer disproportionally from unemployment. If one were to choose a 60 per cent-activity rate as a target, six current EU member states fall far below.[21] Very low female employment often coincides with a sharp decline in older males' employment. Seven EU nations now show activity rates for males, fifty-five to sixty-four, below 40 per cent. The spread, again, is enormous with Portugal, Sweden, the UK and the USA at or above 60 per cent. In some nations early retirement accounts for up to 30 per cent of total pension spending (Thompson, 1998). The single most persuasive policy in favour of sustainability is to raise participation rates, but the logic and consequences of altering older male and female employment is radically different.

4.3.   From work to retirement

In the past, male workers normally retired at age sixty-five, and could expect to enjoy very few years of benefits prior to death. Average retirement age in the EU is below sixty, all-the-while that life expectancy has increased to seventy-four among males, and eighty-one among women. With a longer and healthier life, the average male at age sixty-five will be disability free until about seventy-four and, in the future, even longer.[22] The paradox is that actual retirement age continues to decline.

Were we to progressively raise retirement age up to age seventy by 0.5 years *per annum*, we would be able to maintain pension entitlements at present levels up until year 2040 (see Table 3.8). Given ongoing health improvements, it is not unrealistic to gradually raise retirement towards age seventy. Higher activity rates among older workers may also become necessary in the longer run, since future youth cohorts will be very small indeed.

Retirement is the purchase of leisure against earnings on a life-cycle basis and, omitting poor health and similar concerns, it is a question of preferences and opportunity costs. The incentives system in contemporary pension schemes is, however, strongly biased in favour of early withdrawal. According to Table 3.9, all EU citizens (except in Denmark) have nothing to gain from delaying their retirement decision until, say, sixty-five. A shift

Table 3.8   *Total public pension expenditures (per cent GDP) in 2050*
             according to two scenarios: a baseline scenario of no change,
             and a scenario assuming average retirement at age 70

	Baseline Scenario for 2050[a]	Age 70 retirement scenario for 2050[b]
Austria	14.9	10.6
Belgium	15.1	10.6
Denmark	11.5	8.0
Finland	17.7	11.6
France	14.4	7.6
Germany	17.5	12.8
Ireland	3.0	2.0
Italy	20.3	10.2
Netherlands	11.4	8.6
Portugal	16.5	11.6
Spain	19.1	15.2
Sweden	14.5	7.5
USA	7.0	5.7

Notes: [a] The baseline scenario assumes that present pension rules will remain, a fertility rate of ca. 2 and no change in migration.
           [b] The delayed retirement scenario assumes all of the above, but also that (as of 2005), retirement age will rise by 0.5 years per annum up to a maximum of 70 years.

Source: OECD Income Database.

to actuarial neutrality, alone, would boost participation rates of older workers, and in one case at least, dramatically so. The delay in retirement would obviously be much greater with positive incentives.[23]

Possibly, early retirement is driven more by employer than worker incentives. The motives are many, including the education gap between older and young workers, an issue of importance in an era of rapid technology and skill change.[24] A second factor is wage costs relative to productivity. Seniority-based wage systems, coupled to strong employment protection, give employers huge incentives to utilise the retirement option. If relative wages were more closely linked to productivity, the incentive would probably weaken considerably.[25]

There is, then, a more convincing case for a flexible retirement process rather than slashing pension benefits or coverage. It would be far more equitable if a voluntary decision to retire early were to be transparently connected to a price. As long as a pension system does not penalise workers

Table 3.9   Early retirement incentives and employment among older males: gains from postponing retirement from age fifty-five to sixty-four, and simulated employment effect with actuarial neutrality (1995 data)[a]

	Cumulated pension wealth accruals when postponing retirement from 55 to 64	Simulated increase in participation rate among males, 55–64, with actuarial neutrality
Austria	−3.4	
Belgium	−2.3	
Denmark	0	
Finland	−2.3	+6
France	−1.4	+4
Germany	−1.4	+3
Ireland	−1.4	+4
Italy	−7.9	+20
Netherlands	−1.3	+1
Portugal	−0.4	+1
Spain	−1.4	+3
Sweden	−1.8	+4
UK	−0.5	+1
USA	−1.2	+3

Note: [a] Pension accrual estimates include the effect of saving on pension contributions, and are measured as a multiple of annual earnings. Estimations assume thirty-five years' accumulated social contributions. The simulated change in participation is regression estimated.

Source: OECD.

compelled to retire, say for health reasons, there is no reason whatsoever why workers should not be rewarded for delaying (or gradually sliding into) retirement, even up to age seventy.

Several nations have begun to experiment with flexible retirement, such as part-time work combined with partial pension benefits. Considering, however, the attraction to firms of ridding themselves of expensive older workers, both employees and employers need to be given inducements.[26] 'Life-long learning' solutions or retraining is an alternative for older workers with a substantial skill base to begin with, but is a less realistic (and overly expensive) option if labour markets are flooded with low-skilled older workers, as is often the case today.[27] As more-educated cohorts arrive, such a policy is likely to be gradually more effective. Another option might be to de-couple the wages of older workers from automatic seniority payscales, or to gradually reduce the fixed-cost component of the wage as workers age.[28]

A move towards flexible and later retirement cannot obviously be undertaken overnight, given existing expectations and, too often, the absence of alternatives to either firms or individuals. Indeed, all credible reform proposals emphasise a gradual transition over two or three decades, a transition based primarily on altering incentives in favour of delayed exit without directly penalising those who do opt for earlier withdrawal in the medium term. In fact, OECD's (1998) calculations envisage that a fifty-five-year old worker will gradually increase his/her additional working years from 6.6 in 2005, to eleven years in 2030.[29] Simply put, some thirty years hence we should return to traditional retirement practice, *but* on the new backdrop of much healthier and more-educated older workers employed in much less disabling jobs. The question of altering life-long participation necessarily confronts us with the work–leisure trade-off. How can it be conceived as a welfare gain if policy calls for less life-time leisure and more work? I shall return to this question towards the end.

Redistribution from the aged to the young can be equitable and efficient if it improves the relative welfare of the young without harming the old. Delayed retirement means that the aged contribute more towards their own welfare, but also that the tax and contribution burdens on young workers will decline – a vital issue for raising demand for youth workers. In the final analysis, any intergenerational reallocation of resources and responsibilities – of the kind contemplated here – must be evaluated not only in terms of meeting conflicting welfare goals, but also whether the net result will be an overall efficiency improvement. I shall turn to this aspect later.

4.4. Female employment and the two-earner household

The single most promising ingredient in a long-term win–win strategy is to maximise women's employment possibilities. Besides corresponding to women's own demand for economic independence and careers, the societal gain is to reduce child poverty and to raise the ratio of active contributors to pensioners.

Child poverty is rising, not just because of weakened families, but also because one single income is increasingly insufficient. The erosion of the stable, well-paid, life-long job for males can be, and is, offset by the rising female labour supply which, in turn, is an inevitable development, considering women's higher educational attainment and better job prospects in the evolving services economy. The same goes for female-headed households where mothers' employment is a superior hedge against child poverty. And combating child poverty is not merely an investment in welfare today, but also in the future productive potential of today's children.[30]

To illustrate women's impact on the balance between retirees and workers, let us take two countries with a similar elderly (plus sixty-five) share of total population, namely Italy and Sweden, but with huge differences in female employment. The Italian ratio of elderly to *employed* persons (at 0.52) is almost exactly twice the Swedish (0.27), a huge difference simply due to female employment levels.

5.   THE LINK TO LABOUR MARKETS

The new welfare policy priorities that emerge from the preceding analysis boil down to one basic issue, namely that social policy must maximise citizens' productive resources and life chances. It is important to recognise that any 'work-friendly' policy must align itself to the dynamics of a services-led economy.

The service economy is tendentially dualistic, combining knowledge-intensive professional and technical jobs, with low value added, and labour-intensive servicing. The former are concentrated in business and some social services (teachers, doctors); the latter in sales, consumer and also some social services (restaurant workers, home helpers, nursing assistants). Europe, like North America, is very dynamic as far as business services are concerned. Europe's development of social services is, excepting Scandinavia, sluggish. And European private consumer services are stagnant if not actually in decline.[31]

Contrary to popular belief, services are *everywhere* biased in favour of skilled and good jobs. The dilemma, nonetheless, is that a significant amelioration of mass unemployment means stimulating also low-productivity services, and this means that we must rely also on personal consumer services and social services. The good news is that these are sheltered from international trade competition; the bad news is that they compete directly with unpaid household 'self-servicing'. The problem is that many services are extremely price sensitive. They will grow if, as in the United States, wages and costs are relatively low, and thus affordable or if, as in Scandinavia, they are subsidised by government.[32]

Herein lies the great European policy dilemma. The task of forging a more equitable and efficient social protection system, as outlined above, pales in comparison to the trade-offs involved in stimulating employment-intensive services. *Yet, no solution exists unless we realise that social protection and service employment are directly linked.* The gist of the problem is simple, namely that strong service growth implies more taxation if we emphasise public services or, alternatively, more wage inequality (and lower fixed labour costs) if we emphasise market services.

Most European welfare states and industrial relations systems have committed themselves for decades to a degree of security for the prime-age (male) worker, and a degree of earnings and income equality, that is not compatible with a large, low-end service economy. Moreover, the existing financial pressures on most European welfare states today make it difficult to replicate the Nordic countries' social service expansion twenty years ago.

This problem is now well recognised within EU member states. Witness the extension of targeted wage subsidies (usually aimed at youth and contingent on training), and recent EU-level proposals to stimulate labour-intensive services through a reduction of the VAT.[33] There is virtually universal agreement that strong wage compression, a high tax wedge (especially through mandatory contributions) and, perhaps, overly rigid employment regulation block lower-end services. The great dilemma, though, is that the kind of *tout court*, American-style, de-regulation that would fuel such jobs is unacceptable to European policy makers.[34]

The stagnation of low-end services in Europe is directly linked to the nexus between families and social protection. On one hand, employment-based social insurance systems impose very high fixed labour costs, the marginal effect of which is especially strong in low-wage, low-productivity jobs: a high tax wedge *de facto* prices them out of the market. On the other hand, the lion's share of such service jobs competes with households' own internal servicing capacity. So, where women's employment rate is low, households service themselves; where most women work, households' demand for outservicing increases. In brief, the double-earner families externalise their servicing needs and create jobs.[35]

As discussed above, dual-earner families require services to begin with and herein lies the gist of a win–win policy scenario, namely that more social care services are a key instrument in fighting poverty *and* a potentially very effective employment multiplier. Markets cannot generally guarantee affordable, high-quality care for small children or the aged, and high-quality day care is crucial if our aim is to optimise the life chances of children. In other words, public subsidies or direct public delivery is a basic, first-order priority. An *investment* in women's ability to work is also an investment in family welfare and in job creation. In brief, family services should be regarded not as merely 'passive consumption', but also as active investments which yield a long-run return.

There are bound to be three kinds of sceptics. One would argue that the costs of 'women-friendly' welfare policy, including day care and leave programs, far exceed mothers' marginal productive contribution (see especially Rosen, 1997). The answer depends on how we cost it out. In a 'static' framework, as Rosen adopts, it is undeniably true that the cost of maintaining mothers with small children in the labour market exceeds their contribution. In a dynamic calculus, however, Rosen errs because women who do not interrupt their labour force attachment will lose far less in accumulated life-time earnings than women who do. The cumulative effect of remaining in or leaving the labour market can be dramatic, and this influences mothers' long-term tax contributions. In *Appendix 1*, I provide some estimations which suggest that the public subsidies that mothers receive via day care and paid leave are efficient: the investment pays off because the initial outlays are recuperated and society gains from the additional labour input.

The second source of scepticism comes from those who adhere to the lump-sum-of-labour thesis: work is disappearing. Any glance at comparative statistics tells us this is false, so the question turns to: how many jobs, and what kind? Let us here examine more narrowly the possible net job-effect of women's employment. There is an automatic, first-order effect that comes from providing those child and aged care services that are a precondition for women's careers (and which, most likely, will become female jobs). The issue is whether *net* of publicly subsidised jobs, additional employment effects will be non-trivial. This is a very sticky question. Double-earner families should in principle demand more external services due to higher incomes and due to more severe time constraints. But, again, the issue boils down to relative costs. In Scandinavia where the dual-earner household is now near universal, private consumer services are stagnant (costs are too high). In the United States, however, they are not. A very rough statistical guesstimate suggests that the elasticity is positive and noticeable: possibly, for every hundred mothers who remain employed we shall have an additional net fifteen personal service jobs.[36]

And the third scepticism concerns the emerging low-wage/bad-jobs scenario that inevitably accompanies a large low-end service economy. Since households can resort to time-saving goods (like microwave ovens) or compensate for wives' reduced domestic hours by raising husbands', their need for out-servicing may not translate into effective demand if costs are too high. What, then, is the case in favour of stimulating low-end servicing jobs, either by accepting much greater wage inequalities or by subsidies? It essentially depends on dynamics, namely on the social correlates of low-end service employment.

The social advantages of a low-end labour market are clear: it provides easy-entry jobs for youth, immigrants, low-skilled workers and returning women. Low-end services could play a very positive function *if*, that is, they do not become life-long traps. A brief spell of low earnings and unrewarding work will not, by definition, harm peoples' life chances – to the contrary, if they provide bridges into the labour market or help supplement income. The criteria by which we must judge the costs and benefits of low-end jobs cannot be based on snapshot notions of equality for all, here and now. The only reasonable benchmark is life-course dynamics.

This, in turn, brings us to the preconditions for job and earnings mobility which, research has firmly established, boil down to three main factors: the impact of social origins, of skills and of natural endowments. Investments in child welfare *can* diminish the impact of inherited privilege, of 'social capital', as the most authoritative recent studies conclude (Shavit and Blossfeld, 1993; Erikson and Goldthorpe, 1992; Haveman and Wolfe, 1994). There is less agreement on the kinds of skills that will guarantee decent careers in tomorrow's labour markets, but we *do* know that adequate levels of cognitive abilities may be as important as formal schooling, that social and cultural skills play an ever more important role, and that brief, task-specific training has a far lower payoff than does longer-term, individually tailored training with a theoretical content (Martin, 1998; OECD, 1997). To cite a recent Danish study, the chance of moving out of social marginality increased by 30 percentage points with some vocational training, and by a full 50 points if also accompanied with some theoretical training (Bjorn, 1995). As to endowments, there will always remain a residual population for whom investments in childhood conditions or in later human capital activation will have little effect. No one, I assume, would propose that they be abandoned to a life of poverty.

The point that I shall make in the following, concluding part of this report is that a low-end labour market need *not* be incompatible with a new welfare state scenario. Indeed, one might restate the point in this way: *no win–win welfare model for Europe will be possible unless we accept a different notion of equality*. We have become accustomed to an overly static, here-and-now, concept of redistributive justice: the welfare state must assure that all citizens are protected always. A far more realistic principle would be that our future welfare state accepts, perhaps even sanctions, inequalities here-and-now in order to maximise better life chances for all. If the 'knowledge society' and the modern family create inequalities, the most effective social policy would be one that guarantees that citizens will not become trapped into social exclusion, poverty or marginality across their life course.

6. A EUROPEAN WELFARE STATE FOR THE TWENTY-FIRST CENTURY

Debates on how to construct a new welfare state edifice easily end in paralysis for three reasons. The first is that any reconsideration of the *Gestalt* inevitably raises technical issues related to this or that policy. One quickly loses sight of the broader horizon as soon as one seriously begins to contemplate reforms of pensions, unemployment benefits or home help services for the elderly. It is my understanding that the Portuguese presidency favours a debate on *Gestalt* over technicalities.

The second reason is that 'global' reform scenarios often contain more ideology than relevance. Calling for a privatisation of the welfare state or for a return to family and community is hardly realistic, and therefore not helpful. Neither is, of course, the inevitable rearguard defence of the hard-won *status quo*. The kind of project that the Portuguese presidency has in mind would be poorly served by ideological advocacy.

The third reason may be the most difficult to surmount, namely the short-term imperatives that European politicians and governments face in terms of sustaining existing commitments. If the first priority is to contain expenditures, it would appear a luxury to contemplate new principles of justice and efficiency.

The issue before us has to do with the long term, with the kind of society that our children will live in. And, if this means redefining welfare priorities, we cannot escape the need for some common, basic criterion of what is desirable, given known constraints. What are the common goals to be reached? What do we seek to accomplish? What are the first principles that must guide policy making? What, in brief, can be our common yardstick of justice, of equality, of collective guarantees and individual responsibility? And, once agreed upon, how can our commitments to equity be best put to use in order to maximise efficiency?

6.1.   Basic criteria for policy choice

We must probably assume that most EU countries have reached their maximum limits of public expenditure and taxation. In fact, convergence towards the Maastricht criteria compels expenditure reductions, not bold and expensive reform vistas. The need for restrictive policy already limits the degree to which nations can promote the knowledge society, be it investments in infrastructure, education, training or in improved social welfare.

The resource dilemma worsens considerably when we take into account the new inequalities and social risks that knowledge-based economies inevitably provoke. The evidence is by now clear that the emerging opportunity structure, reward system, and life chances create new winners and losers and, most likely, a deepening gulf between those with skills and those without. The new service economy *can* create jobs, but it *cannot* guarantee good wages and jobs for all. The fabric of our social protection systems will therefore be put to a severe test in terms of nurturing efficiency while securing social cohesion, welfare and equity.

We are compelled to re-prioritise the allocation of our existing welfare package, and this means we must accept at least two ground rules for policy making. One, we cannot pursue too one-dimensionally a 'learning society', a human capital-based strategy in the belief that a tide of education will lift all boats. Such a strategy inevitably leaves the less-endowed behind. Children's ability to make the best out of schooling depends not just on the quality of schools, but also on the social conditions in their families; women are today often more educated than males but will have difficulty putting this to maximum use without generous leave programs and care services.

The second ground rule, following from the first, is that new social policy designs must be primarily financed by a better and more efficient allocation of the resources already available. Entitlement and equity conflicts are easily subdued when the total pie grows. When, instead, we must divide the pie up differently, a clash of interests is hard to avoid. If, then, our task is to identify a superior balance between efficiency and social fairness, we shall be unable to proceed unless we can agree upon a basic, consensual yardstick of justice.

6.2. Decision rules for equity and efficiency

In the pursuit of a 'win–win' strategy, two alternative yardsticks of justice would appear most relevant. The first, and most conventional, is a *Paretian* principle which, crudely put, defines welfare optimisation as any reallocation that produces more efficiency without anyone losing as a result. One example might be wage de-regulation which will surely augment earnings inequalities but, if it creates more jobs, many will benefit and, if the lowest paid are compensated with subsidies, their real disposable income may not decline as a result. The second, and more ambitious, would be a *Rawlsian* principle of justice, namely one which insists that any efficiency gain should be to the greatest advantage of the poorest or weakest. Following the same example, a de-regulatory policy would now be a 'win–win' strategy only if the combination of added jobs and wage effects produce a relative gain among the worst off, and where their gain is to the benefit of all.

Since the EU has, both firmly and explicitly, committed itself to prioritise the fight against social exclusion, the relevant yardstick of justice would appear much closer to the *Rawlsian* principle. Europe, I take it, is not willing to promote the kind of knowledge society which, even if possibly more competitive, will leave large groups behind.

6.3. Principles of reform: towards a social investment bias

Contemporary policy slogans such as work-friendly or women-friendly benefits, life-long learning and social investment strategies, or the popular distinction between active and passive measures, all have in common an implicit distinction between policies that somehow enhance or diminish citizens' self-reliance and capacities, economies' efficiency and productivity. Such slogans reflect a growing unease with the existing bias of compensating the losers of economic change with passive income maintenance, of reducing labour supply, or of parking surplus workers on public benefits. The new policy vocabulary mirrors a growing consensus that social policy must become 'productivist', to coin an expression traditionally used in Swedish policy making. That is, social policy should actively maximise the productive potential of the population so as to minimise its need for, and dependence on, government benefits.

The irony here is that one needs welfare state measures so as to minimise the need for welfare measures. The irony, of course, disappears when we distinguish between types of policy. Some policies can be regarded as an investment in human capabilities and self-reliance; others, while welfare enhancing, are clearly passive income maintenance. Obviously, such a distinction is – and must be – ambiguous. Unemployment benefits appear 'passive', but they *do* aid workers in their search for new employment, and they *do* improve the labour-matching process. Similarly, child allowances add to families' consumption power but they do also diminish poverty and thus enhance children's future life chances. *The important point to stress here is that contemporary policy fashion tends to stress far too narrowly the wonders of 'activation' policies while ignoring income maintenance.* The need for 'passive' measures will not disappear even in the best designed, productivist welfare state: there will always be people and groups that must depend primarily on redistribution, and activating citizens' productive potential often necessitates income subsidies. Regardless, a first principle of any win–win strategy must be that it prioritises social investment over passive maintenance.[37] A second, derivative principle, is that highest priority should go to social investments in children – who are our future productive potential.

7.   TOWARDS A NEW WELFARE STATE DESIGN

If we agree upon a *Rawlsian* principle of welfare optimisation, the preceding analysis points to a set of concrete policy priorities, all perfectly *Rawlsian*:

- maximising mothers' ability to harmonise employment and children;
- encouraging older workers to delay retirement;
- socialising the cost of children mainly by prioritising investments in children and youth;
- redefining the mix of work and leisure across the life cycle;

- reconceptualising 'equality' and basic social rights as being primarily a question of life-chance guarantees.

This will, in the most general terms, imply a greater emphasis on protecting young households, and a stronger emphasis on servicing families.

7.1. The limits of a learning strategy

Accelerating the pace towards a knowledge-intensive economy implies heavy investments in education, training and cognitive abilities. Those with low human and social capital will inevitably fall behind and find themselves marginalised in the job and career structure. It is accordingly tantamount that educational investments be as broad-based as possible. As so much recent research has shown, concrete expertise may be less salient than possessing the essential cognitive abilities required to learn, adapt and be trainable in the first place (see especially OECD, 1997). Activation measures, such as training or retraining, will have a low payoff if workers' initial cognitive capacities are low. They are much more likely to pay off if they are designed around a more comprehensive and individualised 'activation package'.

One pervasive problem across European today is that the stock of low-educated 'excess' workers can be very high – in part because of delayed agricultural decline, in part because of heavy job losses in traditional, low-skill industries, and in part because of an often wide gulf in education between generations of workers. A massive investment in learning will probably reap most of its benefits among younger cohort workers. The dilemma, then, is how to manage the present stock of mostly older, low-skilled males. Early retirement has, so far, been the leading policy and it may have been the only realistic policy so far. Life-long education is an attractive alternative, but may be overly costly and ineffective if the main clientele are older, low-skilled workers. A third policy would be to de-regulate job protection and seniority wage systems so as to align wages closer to productivity differentials – as is generally American practice. This would cause the incomes of youth and older workers to decline, possibly sharply so.

There exists no ready-made formula for a win–win policy in this regard – largely because the problem varies dramatically from country to country. An obvious first step is to assure that future generation workers will have a sufficient skill and cognitive base so that the dilemma eventually evaporates. The problem is the second step, namely what to do with the existing stock. If we are assured that early retirement in the past decades has succeeded in managing what was a transitory glut of elderly low-skilled workers, the dilemma resolves itself and the process of curtailing early retirement can be accelerated. If not, we are left with a mix of continued early retirement, possibly retraining, or downward wage adjustments (or re-employment). The social partners are clearly unwilling to accept across-the-board de-regulation of job security and wages, but it might be an efficiency gain to prolong the employability of older workers by subsidising part of their wage bill. This is especially the case if, as often occurs, retired workers return to work in the undeclared shadow economy. Just like in the case of youth workers, very high fixed labour costs help price them out of the market.

An effective life-long learning strategy can be effective when the basic cognitive skills are already present, and this means that we need to assure that coming generations have the resources required to benefit from investments in training and education across their lifetime. In many EU countries, the existing generational gap is enormous and it is, therefore, tantamount, that this is not reproduced in the future.

7.2. Equitable retirement

The principal problem in today's overly aged-biased welfare systems is that they provide incentives with inequitable results. Workers easily collude with employers to retire early because they will gain little or nothing by postponing exit. At the same time, the pay-as-you-go nature of pension schemes means that retirement at high benefits is heavily financed by the active workforce. Reinstating actuarial incentives to delay retirement would clearly be more equitable and efficient, and it would vastly improve upon the transparency of costs involved in retiring out older workers.

Since workers can expect to be disability-free until age seventy-five, raising and flexibilising retirement age, via incentives, to sixty-five in the medium term, and possibly seventy in the long term, can be positive for individual workers and also for welfare state finances. Abolishing mandatory retirement age and developing flexible mechanisms of gradual exit could be pursued immediately. Longetivity implies that the share of ultra-aged (eighty-plus) just about doubles every two decades. And this means costly and intensive servicing and care needs. If, as I have suggested, retirement households often enjoy 'excess' income and wealth which, if *not* taxed, generates perverse redistribution, an incentive neutral and far more equitable policy would be to earmark taxation of pensioners to their own collective caring needs. Such a taxation mechanism, even if highly progressive, is also likely to be distributionally neutral across pensioner households (the rich generally live longer).

Altering the welfare and work nexus among the aged cannot be an end in itself, but is primarily a means to achieve more inter-generational equity and a more efficient utilisation of public resources. The advantage of an approach, like the one I have sketched out above, is that it is demonstrably far more *Rawlsian* (and, in fact, also *Paretian*) than any alternative. There are really only two genuinely effective policies to combat the long-term financial consequences of ageing: sharply reducing pension entitlements, or raising participation levels. Reducing entitlements means stimulating private pension plans for large parts of the population. The problem with a private-dominated pension mix is that it replicates life-course inequalities, and the more that private plans grow, the more is it likely that we shall see downward pressures on public benefits targeted to low-income households. Even if a system dominated by private schemes augments national savings rates (and thus 'efficiency'), they will possibly lead to non-Paretian outcomes: the weakest may end up worse off. Identifiable trends in labour markets also threaten the viability of a predominantly private pension structure since declining job security and more intense inequalities will negatively affect workers' capacity to accumulate individual savings.

7.3. Harmonising family welfare and labour markets

Post-industrial, service-dominated labour markets cannot avoid producing new inequalities. One source of dualisms comes from systems of strong protection of stably employed 'insiders' with a possibly growing clientele of 'outsiders', such as precariously employed temporary workers or the unemployed. Insider–outsider cleavages tend to affect youth and women workers most negatively. A second source of new inequalities comes from the rising relative wage premium of skills. And a third will emerge if, and once, labour-intensive consumer services grow. The standard trade-off between jobs and inequality, epitomised by the US–Europe comparison, is far too simplistic, but it is difficult to imagine a return to full employment in Europe unless also low-paid and often low-productivity service jobs are encouraged.

European industrial relations systems and welfare policy are generally premised on a commitment to wage equality and job security. Hence wage minima, contractual regulations, and high fixed labour costs are difficult to touch. There are two prevailing arguments against a 'low-wage labour market' through de-regulation. The first, and most convincing, is that US-style de-regulation not only creates huge inequalities, but also threatens the basic fabric of trust and cooperation built into European models of social partnership. Europe's tradition of broadly negotiated 'efficiency wage' arrangements is, to a great degree, its comparative advantage. The second, and far less convincing, argument is that a low-wage service economy poses a direct threat to families. The defence of existing regulatory practice is often premised on the traditional assumption that families' welfare depends almost exclusively on the wages, job security, and accumulated entitlements of the male breadwinner.

This family model is in rapid extinction. Unfortunately, some of its latter-day successors, like single-parent households, are at high risk but much less so if the parent is able to work. Two-income families enlarge the tax base and minimise the welfare lacunae that prevail when wives' entitlements are derived from the husband. And the dual-earner family is the single best strategy to minimise child poverty. Two workers are, moreover, an effective household buffer in the eventuality of employment interruptions. It follows that a strategy premised on investments in education *must* be coupled to a recast family policy, the cornerstones of which must be guarantees against child poverty. Such guarantees must centre on affordable child and aged care, on adequate child benefits, and on maternity and parental leave arrangements that minimise mothers' employment disruption and maximise their incentive to have children. In the long run, therefore, the most persuasive 'win–win' strategy is to re-direct resources to child families if our goal is to sustain our long-term welfare obligations towards the aged while effectively combating social exclusion.

Whether the externalisation of family care is placed in the market or directly furnished by public agencies is unimportant, as long as standards and affordability are guaranteed. Were we to pursue an explicitly *Rawlsian* family policy, there would be a strong case for prioritising high standard child care services to the weakest families since optimal quality care may offset inequalities that stem from uneven social capital within families.

7.4. Life chance guarantees rather than here-and-now equality for all

Any assessment of the pros and cons of heightened labour market inequalities must be premised on a dynamic, life-chances perspective and not, as is typically the case, on a static view of fairness and equality. Low-paid jobs, even at near-poverty wages, are not by definition a welfare problem. The acid test of egalitarianism and justice is not whether such jobs exist or what share of a population is, at any given moment, low paid. Low-end employment would be compatible with '*Rawlsian*' optimisation if it does not affect negatively peoples' life chances. The issue here is entrapment and mobility chances.

On this score, unfortunately, research has not yet provided much indisputable evidence. We do know that a sizable minority of American low-wage workers remain trapped for many years (a higher rate than is the case in Europe where, comparably, low-wage jobs are much rarer). And those most likely to become trapped are the low skilled. We also have fairly good evidence that, net of family origins, skills and education constitute the single best guarantee of mobility. Hence, expanding low-wage service jobs in tandem with heavy investment in skills would, for the majority, constitute a win–win policy. The problem lies in the risks of entrapment among the minority which, perhaps, is 'untrainable' or, for various other reasons excluded from mobility. It is precisely for this reason that a 'learning strategy' needs to be accompanied with a basic income guarantee strategy.[38] Nonetheless, the problem of inequality would disappear if the welfare state extends a basic life-chances guarantee to citizens: a guarantee of job mobility via education or, alternatively, a guarantee that condemnation to a life of low wages does not imply income poverty throughout the life course.[39]

At the risk of repetition, the greater is the investment in social resources among children, the greater will be the later payoff in terms of life-long learning abilities and readjustment, and the smaller will be the burden of compensating the 'losers'.

8.   LEISURE AND WORK

The kind of 'win–win' scenario presented above appears heavily biased towards work. Notwithstanding sluggish growth, European GDP *per capita* is now 50 per cent greater than before the 1970s oil shocks. And such wealth ought to translate into more leisure time. Clearly, no *Rawlsian* scenario is possible without a convincing balance between leisure and work. On this count prevailing thinking is extremely muddled, often maximalistic, combining both a static and dynamic perspective.

Contemporary European debate is dominated by the controversy over the thirty-five-hour week, promoted for its purported positive effects on job creation. If, indeed, its main goal is to stimulate employment the strategy is, at best, controversial and, at worst, self-defeating. If the goal is to extend leisure time, the question that few have posed is, why focus on weekly or monthly, rather than on life-time distributions of leisure and work?

The irony is that the call for a shorter work week follows several decades of significant work-time reductions on a yearly and a life-cycle basis. The typical EU country's annual working hours is now down to 1600–1700 hours, mainly attributable to the spread of part-time work, vacations, holidays, and paid leave arrangements, and – unfortunately – also to unemployment and exclusion from the labour market. Much more dramatic are reductions in lifetime employment. The average (male) worker in 1960 would work for roughly forty-five years; his contemporary equivalent will work perhaps thirty-five years. It is not altogether clear to what extent more leisure is voluntary and desired, and to what extent it reflects inability to attain gainful employment. In the case of women, leisure is often unpaid housework.

Should we favour more leisure on all accounts? Fewer weekly hours, annual hours *and* life-time hours? If so, do we agree on the associated economic opportunity cost? Is it equitable if the cost of leisure for some is shifted onto the shoulders of others? Do our leisure-time arrangements adequately maximise our productive potential? Can we envision alternative, more equitable and efficient, distributions between leisure and work? These are questions that almost no one raises in the current social policy debate, but they are crucial for any consideration of a new welfare order.

To some extent, prevailing leisure–work arrangements are intended, such as maternity or parental leave. But, to a large extent, they are due to unintended consequences of policies designed to (or unable to) solve completely different problems. Early retirement and unemployment are obvious examples. Leaving aside 'unwanted' leisure, do we in fact have an adequate understanding of what would be citizens' optimal leisure–work preferences? I think not. Early retirees may be individually content to exit prematurely, given constraints and incentives. But if these incentives are societally harmful and were thus removed, would the desire to exit remain? Would Italian women's employment rates follow Holland's if restrictions against part-time work were eased? Or would Dutch women's working hours approach full time if, as in Scandinavia, affordable day-care were available?

The chief problem is that past policy has resulted in overly rigid leisure–work arrangements that permit individual workers little choice as to how to optimise their own mix. At the same time, work–leisure incentives for some groups are gained at the cost to others. Basically, existing practice reflects a social order that is no longer dominant. We have, so far, bundled free time within the working week, within stipulated vacations and holidays, and at the tail end of our lives. If our goal is to optimise life chances in a dynamic sense, such an order may not be compatible with the exigencies of an evolving knowledge society.

Emerging trends in family and labour market behaviour suggest that citizens' demand for leisure and work may be spaced out across the life cycle in a radically different manner than so far. The case of paid maternity and parental leave is one of the few examples of policy that seeks to address emerging incompatibilities. A full-fledged 'life-long learning' model will require similar arrangements, namely paid education or training leaves. There is a strong case to be made in favour of the idea floated within Nordic social democracy in the 1970s (and recently re-launched by Claus Offe), to re-think the work–leisure mix in terms of lifetime 'sabbatical accounts'. The idea is that citizens (after a minimum number of contribution years) can draw upon their retirement savings accounts at will, be it for purposes of education, family care or pure vacation. There is, in principle, no reason why retirement should be concentrated in older ages. A radical version of the win–win scenarios developed above would, in fact, call for the abolition of retirement as we know it and redefine it as an issue of pacing individuals' life course. If some are more risk adverse, they will opt for educational leaves or minimal career interruptions; if some are more leisure prone, they will favour interruptions. The bottom line is that citizens have much greater individual command over how to design their own life course, over how to mix work, education, family and free time. If the financial consequences are transparent, an individual will be able to rationally decide whether the choice of time off at age thirty-five against one less year of retirement is advantageous or not.

A POSTSCRIPT

The Portuguese presidency was a clear success in terms of pushing forward a common European commitment to improving its welfare performance. There appears to be broad agreement behind what we might call a minimalist skeleton for a new social model. As summarised by the European Council, the key words are more investment in people, activating social policies and action against social exclusion. The formulation of these objectives is, of course, quite vague and intrinsically uncontroversial. The European Council documents do, however, also elaborate on some concrete steps to be taken and on some targets to be achieved. There is a call for service sector job growth, for delayed retirement, more female employment and for servicing families. In some cases, the targets are quite precise: by 2010, unemployment should drop to 4 per cent, overall employment levels should reach 70 per cent and female employment, 60 per cent; the poverty rate should drop from 18 per cent today, to 10 per cent; investments in human resources should increase by 50 per cent; and the number of youth with less than secondary level education should be halved.

That the member states are seriously committed to a set of more concrete social objectives emerges also from the French presidency in the autumn of 2000. This is especially clear as regards retirement policy objectives (which, like this report, upholds the need for strong pension guarantees while calling for delayed retirement) and measures to combat poverty and social exclusion (which agrees with this report in its call for more employment promotion, for concerted policy to alleviate women's double burden, and for the elimination of 'social exclusion' among children).

If we view the Portuguese (and now the French) Presidency as a first step towards an ambitious project of building a new welfare state edifice, what should follow? How might we assure that real practical progress is made during the coming decade? To further such a project, I see two or three issues that call for immediate follow-up. One has to do with *social account-ing*, an issue briefly mentioned in my report. This issue, in fact, embraces two dimensions. Firstly, we require a common system of indicators to monitor and define social progress and relevant targets. Certainly, we already have available a fairly large battery of good indicators for employ-ment and the labour market, for incomes and poverty, and the like. It now appears that the EU has adopted the median as a common baseline measure for poverty – a first basic step towards comparability. But, if one were to follow the gist of the analysis provided in my report, the really important measures of welfare have to do with life-course dynamics, with citizens' chances of welfare improvements or, possibly, their risks of entrap-ment. To give an example drawn from my own policy priority menu, child poverty needs to be eradicated so as to maximise children's school perfor-mance, cognitive development and overall life chances. How would we design a policy for this aim? The answer depends on a clear understanding of how, precisely, do poverty spells affect children. Do negative conse-quences kick in if the duration of poverty exceeds three months, six months or one year? What are the mechanisms at work? Is income poverty a measure that adequately summarises the multi-dimensionality of poverty? A basic precondition for effective policy in this, and in most similar domains, is that we possess good information on the *dynamics* of peoples' welfare. As with child poverty, it appears very difficult to design a policy for delaying and flexibilising retirement unless we know more about who is most prone to exiting early, why, and under which conditions. Again, we lack the kind of comparable, European-wide data that inform us about such dynamics. It is urgent that the EU take serious action to (re-) launch a credible European Household Panel.

Secondly, our existing systems of income and expenditure accounts do not appear particularly relevant for the kind of 'new social model' that we are working towards. Some existing accounting problems are well known, such as the necessity of distinguishing between gross and net social spend-ing. It is more problematic that public social expenditures are routinely defined as current outlays and consumption (unlike airports, which are classified as capital investment). Nonetheless, as also many of the new popular policy 'buzzwords' signal, our desired welfare state will put its accent on 'investing in people', 'activating human resources', and the like. From a legitimacy point of view, as well as from a practical accounting view, it is paramount that we understand correctly the nature of social outlays. The costing-out example of day care and parental leave, provided in the Appendix, reveals part of the problem: current outlays to working mothers may appear extravagantly expensive here and now, but if they are recuperated in the long haul because of improved female earnings capacity, the same outlays have a rate of return (both private and social) that can be estimated. In such a case, the policy issue is not whether or not to provide for working mothers, but rather: can we envisage a policy that similarly helps mothers and which can yield an even better rate of return?

A second, and also quite urgent, follow-up entails a much more complete cataloguing of the emerging risk and needs structure. It is, by definition, quite difficult to identify emerging realities in an epoch of radical socioeconomic change. Our situation is akin to the driver in dense fog: we can still clearly envisage the place we left, but have great difficulty seeing whatever lies ahead. To be sure, the fog is lifting partially and a number of emerging risks are already quite evident: low-skilled workers and the lone-parent family are examples. Other emerging risks and needs appear on the horizon, but with less clarity: the welfare problems associated with longevity, for example. And still others may, as yet, be very difficult to identify. As an illustration, let us imagine the society that will ensue when women *de facto* match males' employment levels. One possible result is that neighbourhoods, or civil society, empty out during daytime: both parents are at work, and the children in care or school. What might be the social welfare consequences of such a model? Will it, as Arlie Hochchild argues, result in a perverse world in which our place of work becomes also our centre of social fulfilment? Will it produce even stronger mechanisms of marginalisation for those who remain excluded from work or education? Could such a scenario be averted by policy? Or take another, little understood, trend – the apparent rise of 'no-work households'. We know very little about long-run trends: is it a growing phenomenon? Nor do we know much about who is most at risk, or why. Are there trends under way in marriage and cohabitation patterns that widen the gulf between winners and losers?

Such an effort to better catalogue new risks must be connected to the existing social policy repertoire of Europe's welfare states. Our overriding challenge, as stated, is to construct a new social model. A first necessary step in this direction is to have a clearer idea of where exactly exist lacunae or disequilibria between welfare demand and supply.

In closing, permit me to make yet one more appeal for a strong family and child-centred strategy of welfare state reconstruction. A recast social model means, by definition, a future-oriented project and, as such, it should begin with those who will be the adults of that future. When we also define our goals for that future in terms of maximising Europe's competitive position in the world economy, it is obvious that this entails investing in today's children. The European Council's stated goal of 'more investment in people' is uncontroversial until the moment when investment funds must be allocated between young, prime-age or elderly 'people'. It is a formulation that skirts the thorny, but essential, issue of how should we define our priorities if, as is likely, we cannot make everybody equally happy. The issue is indeed thorny because of vested interests and the balance of political power. Europe's median voter is getting older, and the electoral balance shifts evermore in favour of the aged. In addition, interest organisations mainly represent the 'insiders', those with stable jobs or strong resources. Hence, everyday politics favour those who, most probably, prefer to maintain the welfare state status quo rather than a reprioritisation in favour of those at risk. My appeal can be interpreted as yet another suggestion for a follow-up to the Portuguese presidency, namely the need to concretise goals and targets by more clearly selecting among evidently competing priorities. The European Union and member state politicians need for this purpose to decide upon *and* consequently propagate a decision rule of justice and fairness that will permit concrete prioritisation and consensual implementation of a social model truly dedicated to social cohesion ten, twenty or fifty years hence.

APPENDIX 1   COSTING OUT SOCIAL INVESTMENTS IN FAMILIES

Women's integration in the labour force increasingly means a life-long commitment to paid employment, and this poses a basic trade-off to families between careers and children. Bluntly put, new emerging family types are no longer willing to internalise the cost of having children in light of the opportunity cost involved in women's careers or household necessity of doubling up earnings.

Supporting working mothers implies a combination of adequate and affordable daycare for ages 0–6, and income maintenance during maternity and parental leave, as well as covering predicted work absence when children are ill. The heavy spending involved has provoked debate about the cost effectiveness of such policy. Rosen (1996), pointing to mothers' sharply reduced hours while children are small, argues that it constitutes a substantial resource waste, an exorbitant subsidy of 'negative productivity'. Critics against this view hold that the transitory cost of subsidising mothers' labour force attachment should be judged in dynamic terms, evaluating its effect on life-time earnings accumulation and certainly tax payments to government.

A full estimation of the 'real' public cost necessitates data not easily available, particularly regarding mothers' real productivity while they receive public support, and their expected lifetime earnings. As to the latter, we can use recent benchmark estimates which suggest an accumulated lifetime earnings loss of 2 percentage points per year if the mother exits the labour force for five years; if, instead, she remains on part-time during the same period, the loss drops to only 0.5 per cent (Bruyn-Hundt, 1996). Thus, if the mother can expect to work another twenty-five years, the cumulative effect of full interruption can be dramatic (50 per cent), while modest in the case of part-time continuity (12.5 per cent). This clearly makes a big effect in terms of associated tax contributions.

I utilise Denmark as a costing-out example. Let us begin with an overview of public costs per child (1995 data):[40]

- per child daycare cost, ages 0–3:   DKr. 41.945
- per child daycare cost, ages 0–6:   DKr. 57.195
- gross cost of parental leave: DKr. 92.638
- total cost per child, age 0–6, if mother is on leave[41]   DKr.100.938
- total cost per child, age 0–6, mother uses daycare   DKr. 65.495

The annual per child cost of supporting working mothers is clearly steeper during parental leave, but since benefits are taxed, the *net* cost is only DKr. 62.000. The cost of day care instead is lower, and in net terms much lower because the mother would be working and paying taxes. Assuming the mother earns 67 per cent of an average wage, she will be paying taxes equivalent to DKr. 63.400 – thus more than the daycare subsidy per child.

We should assume that the average mother has two children, and that she will make use of day care over a period of, say, five years at an annual cost to government of Dkr.114.000 per year (which is roughly twice the taxes she will contribute during the period if she works). The net outlays by government for the entire five year period is, then, Dkr. 250.000. We cannot include 'hidden' additional costs, such as absence from work when a child is ill (here the net cost would be Dkr. 245 per day).

In a dynamic, lifetime earnings framework, our estimates of the cost effectiveness of childcare becomes quite favourable. Let us first take the mother (two children) who interrupts *fully* for the same five years, and let us conservatively put her per annum earnings loss equal at 1.5 per cent. She would, over a subsequent twenty-five year career experience a 35 per cent cumulative income loss. At a 67 per cent earnings level, this would in constant Kroner total to Dkr. 1.400.586. The associated revenue loss to government (assuming no change in taxation rates) would amount to Dkr. 490.205.[42] Put differently, if a mother does not interrupt her career, she will contribute an additional Dkr. 490.000 (or just about *twice* of what child care originally cost) in tax revenue. To this estimate, however, we must also add the public cost associated with two years' paid parental leave (one year for each child) which costs the government *net* of taxes Dkr. 124.000. In this case, too, the government comes out ahead with an additional Dkr. 115.000 gained over the mother's life career.

The above costing-out exercise is certainly highly stylised, and the net gain of Dkr. 115.000 to the tax authorities might look somewhat smaller if we could include costs of work absenteeism. For example, were the mother to take thirty days of absence over the five-year period, we should deduct roughly Dkr. 7350. On the other side of the coin, the mother's sustained work attachment and earnings imply much lower risks of household poverty (and thus less likelihood of drawing upon public income support).

NOTES

1. I refer to Robert Lindley's chapter in this book for a more thorough discussion of the issues involved.
2. This view is equally fallacious (and quite similar) to the classical marxist idea that social-isation of the economy would solve most social problems, or to the post-war meritoc-racy argument which believed that mass education would eliminate class differences.
3. Here we might add yet another policy priority: re-think our existing accounting systems so as to recognise that many social outlays are direct or indirect *investments* with a cal-culable economic payoff.
4. This paragraph draws on Esping-Andersen (1999). International research and policy making has, in recent years, awoken to the necessity of examining welfare problems in this way. See, for example, OECD (1997; 1998).
5. For details, see Bison and Esping-Andersen (2000).
6. Note that there is overlap between the two
7. Social transfers account for only one-third of working single mothers' total income in Scandinavia. For details, see Gornick Meyers and Ross *et al.* (1997) and Esping-Andersen (1999).
8. The aged bias of social expenditure has *increased* in seven out of thirteen EU member states since 1980.
9. My own estimates suggest that due to high fixed labour costs and wage compression, full-time, full-year day care in countries such as Germany or Italy costs about half of what an average full-time employed mother can expect to earn. A significant reduction of rel-ative servicing costs can only realistically occur against the backdrop of a radical de-regulation of wages and reduction of fixed labour costs.
10. We usually define the poverty line as less than 50 per cent of median (adjusted) dispos-able income
11 Public transfers account for the lion's share of total disposable income in countries like France, Germany and Sweden (70–90 per cent), but far less in others (such as the UK or the USA, where private pension plans and accumulated savings play an important role) (for data, see OECD, 1998). Earnings (often undeclared) can play an important role in pensioner income packages. This may be especially pronounced in cases, such as Italy, where early retirement is prevalent and where there exist strong incentives to supply and demand workers who do not need to pay fixed labour costs. Pension schemes are, in some cases, clearly subsidising the informal economy.
12 In some countries, young families' access to housing depends heavily on inter-generational capital transfers of this kind.
13 To illustrate the point, workers at age (ca) 60 earn 100 per cent of average wage in Denmark and the UK, a full 140 per cent in France, but only 80 per cent in the USA:

Estimated age-wage relativities for males. 1990. *(Average = 100)*

	At age ca. 20–25	At age ca. 50	At age ca. 60
Denmark	85	105	100
France	70	120	140
UK	80	125	100
USA	65	105	80

Source: OECD (1996; Chart 4.3).

14 A recent US study finds that employees in non-standard employment (both at the high and low end) are far less likely to have private pension savings (Economic Policy Institute, 1997).

15 The OECD (1997) estimates that workers with less than secondary education can expect five to seven years' of unemployment over their lifetime in the UK, Finland and Spain, and between three and five years' in Ireland, Germany, Sweden, France, Belgium, Denmark and Canada.

16 The OECD (1997) study cited above indicates that expected unemployment years across the life course for people with post-secondary education is less than half that for low educated.

17 The same argument holds for privatising pensions. Just like public insurance schemes, private plans work well for workers with long, stable and well-paid careers. Coverage is low among employees in atypical (such as part-time or temporary) employment, and traditional employer occupational pension plans are eroding as a result of the decline of large firms. Encouraging private pensions for the top half of the labour market and limiting public pension commitments to the bottom half of the population is certainly one possible long-term scenario. I assume, however, that such a scenario is not on the political agenda in the large majority of EU countries. Targeting public pensions only to the poor would reduce the public expenditure burden dramatically, but to put it bluntly: why should we construct inequalities in the future when it is not necessary? Privatisation will never qualify as a Paretian welfare improvement. As far as taxing retirement income is concerned, one should clearly avoid too much taxation since this may produce negative savings incentives among pre-retirement workers. If there is inequity in the distribution of resources between the aged and the young, a system of taxing excess incomes among the aged would be acceptable (and more incentive-neutral) if it were earmarked to cover other risks among the elderly (such as disabilities and intensive care needs).

18 Forecast simulations suggest that a move to strictly targeted public pensions (covering the bottom third only) would bring most countries' pension finances into balance by 2050 (see OECD, 1996).

19 It is possible that the net effect of encouraging more children would be stronger in an indirect sense since all indications are that very low fertility in Southern and Continental Europe (but *not* in Scandinavia) is concentrated among the most educated women.

20 Further decline should not be ruled out. Present Italian and Spanish fertility stands at about 1.2, but Lombardia and Galicia now hover around 0.8! Golini's (1994: 54) long-term demographic projections suggest that if Italy's rate of 1.2 continues throughout the twenty-first century, the Italian population will decline to less than half its present level by 2092.

21 Such a target norm corresponds to the average between EU and North American rates

22 At age sixty-five, French men can expect ten years free of moderate disabilities; Germans a full twelve years (Jacobzone *et al.*, 1998)

23 The early retirement incentive is, at least in some EU countries, additionally problematic if retired workers return immediately to work in the informal economy at sharply reduced real wage rates (in Italy, where this is widespread, fixed labour costs approximate 50 per cent of the total wage bill). Although hard data are impossible to come by, there

are ample reasons to believe that early retirement has become an invisible wage subsidy to older workers, one that perversely augments tax avoidance and black market activity.

24 The education gap is generally very large between current retirement age workers and the twenty-five to thirty-four age group, but the gap begins to disappear already with those in the forty-five to fifty-four age range. In other words, we might expect that the education-deficit effect is a present, transitory one, and that it will lessen considerably over the next ten to twenty years (based on author's calculations from the OECD Education Database).

25 Such a degree of wage deregulation is, however, difficult to imagine in the context of most EU member countries' industrial relations systems. Here one must also consider the welfare consequences of gradual earnings decline, but this effect may be more neutral than what appears at first glance since the alternative, early retirement, *also* produces real income decline. As an alternative policy, one might consider indirect subsidies such as a gradual reduction in the fixed labour cost component of older workers' pay.

26 Part-time employment among males, sixty to sixty-four, is high in Austria, the Netherlands and Sweden (about 35 per cent), and also substantial in Denmark, Finland and the UK. It is virtually non-existent in Greece, Ireland, Italy and Spain (European Union Labour Force Survey, 1995).

27 In fact, workers older than fifty-five are half as likely to participate in adult education as are twenty-five to thirty-four year olds (with Sweden as a major exception). Also, adult training and education is twice as likely among better-educated workers.

28 There is some, albeit old, evidence to suggest that a decrease in earnings as workers age is viewed as legitimate (Rainwater, 1974)

29 Although any reform of retirement age must be implemented gradually, there is no reason not to abolish *compulsory* retirement at age sixty-five immediately.

30 The empirical evidence on the effects of low incomes in childhood on later achievement in school and careers is, by now, very strong. See, for example, Duncan and Brooks-Gunn (eds) (1997), Haveman and Wolfe (1994), and Mayer (1997). Simular results come from the Swedish Level of Living surveys.

31 The following is based on Esping-Andersen (1999)

32 In the Nordic countries, up to a third of total employment is in the public sector, fuelled by social service growth. There, like across the European continent, private consumer services are generally 'priced out of the market' – indeed, they have been declining over the past decades.

33 Individual countries, like Denmark, have experimented with alternative subsidisation schemes to induce more consumption of service labour. Often such subsidies are an attempt to avoid that lower-end services end up in the black economy.

34 And, such de-regulation would almost surely have adverse consequences across the entire labour market, not to mention that it would by necessity imply a major roll-back of existing welfare guarantees.

35 Hence, women's average weekly hours of unpaid domestic labour is almost twice as high in Spain as in Denmark.

36 This estimate is based on time series analyses for the US, France and Spain. For details, see Esping-Andersen (1999; 118).

37 Contemporary national accounts systems are unable to distinguish between social expenditures that play an 'investment role' and those that do not. Parallel to the distinction between capital and consumption accounts, some social expenditures arguably enhance a nation's capital stock and reap a dividend. The actual task is daunting and full of ambiguities, but this is also the case in conventional national economic accounts (should a tank or a jeep for the military be classified as investment or consumption?).

38 Whether such an income guarantee is designed around the Anglo-Saxon formula of work-conditional income supplements or along more traditional social assistance lines is left open.

39 It is very important to distinguish this 'life-chances' guarantee from conventional 'guaranteed citizen income' plans that many advocate. Above all, the life-chances guarantee is meant to be premised on work and not, like the latter, on the assumption that there

will not be sufficient work available. Indeed, the main principle here is to reward the incentive to work. This is not the place to discuss the practical design of such life-chance guarantees. Clearly, active training and learning policies will come to play a core role. One might consider a variant of the American 'earned income credit' subsidy, or similar 'negative income tax' models, as a means to guarantee welfare for those who end up trapped in inferior employment.

40 Calculations of parental leave benefits assume maximum benefit (70 per cent replacement) for one year. Daycare costs are estimated by dividing public day care spending by number children covered. Source: *Social Security in the Nordic Countries* (1996).
41 Including standard child allowance payment
42 These are very conservative estimates since they do not take into account the likely earnings increases that she will have through her career.

REFERENCES

Bertola, G., J. Jimeno, R. Marimon and C. Pissaridis (1999), 'Welfare systems and labour markets in Europe: what convergence before and after EMU?', Paper prepared for the DeBenedetti Foundation Symposium, Genova, July.

Bison, I. and G. Esping-Andersen (2000), 'Income packaging, poverty and unemployment in Europe', in D. Gallie and S. Paugham (eds), *The Experience of Unemployment in Europe*, Oxford: Oxford University Press.

Bjorn, L. (1995), 'Causes and Consequences of Persistent Unemployment', Ph.D. dissertation, Department of Economics, Copenhagen University.

Bruyn-Hundt, M. (1996), *The Economics of Unpaid Work*, Amsterdam: Thesis Publishers.

Duncan, G. and Brooks-Gunn, J. (eds) (1997), *Consequences of Growing up Poor*, New York: Russell Sage.

Easterlin, R. (1987), 'The new age structure of poverty in America', *Population and Development Review,* 13 (2), 195–208.

Economic Policy Institute (1997), *Managing Work and Family: Nonstandard Work Arrangements among Managers and Professionals*, Washington DC: Economic Policy Institute.

Erikson, R. and J. Goldthorpe (1992), *The Constant Flux*, Oxford: Clarendon Press.

Esping-Andersen, G. (1999), *Social Foundations of Postindustrial Economies*, Oxford: Oxford University Press.

Golini, A. (1994), *Tendenze Demografiche e Politiche per la Populazione. Terzo Rapporto IRP*, Bologna: Il Mulino.

Gornick, J., M. Meyers and K. Ross (1997), 'Supporting the employment of mothers', *Journal of European Social Policy*, 7, 45–70.

Haveman, R. and B. Wolfe (1994), *Succeeding Generations. On the Effects of Investments in Children*, New York: Russell Sage Foundation.

Jacobzone, P. *et al.* (1998), 'The health of older persons in OECD countries: is it improving fast enough to compensate for population ageing?', *OECD Labour Market and Social Policy Occasional Papers*, no. 37.

Kohli, M. (1999), 'Private and public transfers between generations', *European Societies*, 1(1), 81–104.

Martin, J. (1998), 'What works among active labour market policies', *OECD Labour Market and Social Policy Occasional Papers*, no. 35.

Mayer, S. (1997), *What Money Can't Buy*, Cambridge, MA: Harvard University Press.

Mirowsky, J. and C. Ross (1999), 'Economic hardship across the life course', *American Sociological Review*, 64 (4), 548–69.

OECD (1996), *Ageing in OECD Countries*, Paris: OECD.

OECD (1997), *Literacy Skills for the Knowledge Society*, Paris: OECD.

OECD (1998), *Maintaining Prosperity in an Aging Society*, Paris: OECD.

OECD (1999), *A Caring World. The New Social Policy Agenda*, Paris: OECD.

Orszag, P. and J. Stiglitz (1999), 'Rethinking pension reform: Ten myths about social security systems', Paper presented at the conference on New Ideas about Old Age Security, The World Bank, 14–15 September.

Palmer. J., T. Smeeding and B. Torrey (1988), *The Vulnerable*, Washington DC: The Urban Institute Press.

Preston, S. (1984), 'Children and the elderly in the US', *Scientific American*, 251, 44–9.

Rainwater, L. (1974), *What Money Buys: Inequality and the Social Meaning of Income*, New York: Basic Books.

Rosen, S. (1996). 'Public Employment and the Welfare State in Sweden', *Journal of Economic Literature*, 34: 729–40.

Scharpf, F. (1999), *Governare l'Europa*, Bologna: Il Mulino.

Shavit, Y. and H.P. Blossfeld (1993), *Persistent Inequality*, Boulder, CO: Westview Press.

Social Security in the Nordic Countries, 1996. Copenhagen: Nov.

Storesletter, K. (2000), 'Sustaining fiscal policy through immigration', *Journal of Political Economy*, 108 (21), 300–23.

Thompson, L. (1998), *Older and Wiser: The Economics of Public Pensions*, Washington DC: The Urban Institute.

4. Knowledge-based economies: the European employment debate in a new context

Robert M. Lindley

INTRODUCTION

Knowledge is the basis of much behaviour: the search for and exploitation of it have been at the heart of social and economic development for centuries. Yet there are now claims that radical changes are afoot which will greatly increase the significance of and alter the pattern of knowledge production, dissemination and use. Countries and organisations that understand this and adapt to take advantage of the enormous opportunities in prospect will, it is argued, place themselves in strong positions to compete effectively in the global economy. 'Knowledge workers' will emerge as the dominant occupational group with high levels of education, continuing professional development and autonomy. They will be the first to connect to the evolving global community – many already are.

Social exclusion occurs when a society fails to organise itself to ensure that all its members can participate. Those excluded suffer from a cumulative disadvantage which goes well beyond their individual characteristics and experience and extends to the local communities in which they live. There are personal, family and wider collective ingredients to the process of their exclusion. They will be the last to 'get connected' electronically as well as socially.

Knowledge workers and the socially excluded seem destined to live in different worlds. But where will the rest of the population who fall into neither group live and work? What are the mechanisms by which the knowledge-based scenario might be moulded to meet social as well as economic objectives? How distinctively different is this scenario in any case? Have management fad and commercial hype led to the coining of a new pseudo-intellectual currency? Are governments caught up in the global transmission of shallow thinking before having time to absorb the lessons of the recent past? If there really is something substantial amongst the speculation and rhetoric, how does it alter the balance of judgement about the feasibility and desirability of presently debated policy options?

This contribution explores some of the aspects of such key issues raised by considering the development of a knowledge(-based) economy or society. The focus is mainly on the potential consequences for employment as far as they can be discerned and on policy in the labour and learning fields. However, ideas relating to the knowledge society have a certain edge to them that goes beyond the socioeconomic scenario they seek to characterise. They are heavy with implications for the *behaviour of the policy development system itself.* Perhaps it is this, as much as the substantive change allegedly under way, that calls for a sharper perspective on the European employment debate.

The remainder of this chapter is in five main sections. Section 2 deals with the overall nature of the changes envisaged which make an appreciation of and responses to the potential implications of the knowledge economy appear so important. Section 3 examines the employment trends identified in some leading projection exercises and explores the key insights available from more qualitative studies. It then turns to the analysis of the ways in which the knowledge-based economy might reinforce or modify some of the underlying forces that are affecting the evolution of job structures and content. Finally, this section considers the forms of adaptation necessary at the levels of the 'organisation' and the 'economy', relating the discussion to the kinds of mechanism thought likely to promote high-performance enterprises, especially as they relate to knowledge production and transfer, initial education and training and continuing professional development.

Section 4 turns to address the implications of the knowledge economy for the 'other world' of social exclusion, starting by elaborating what is meant by social exclusion and going on to see how the evolving knowledge economy could make it more difficult to achieve the frequently stated but modestly realised aims in this area of policy. It then looks at how wider participation might be accommodated within a 'knowledge *society'.*

Section 5 turns to the conduct of the policy community itself as it increasingly promotes the overall vision of the knowledge society and begins to reflect more rigorously about how its own role should be conceived and carried out. Section 6 summarises the main conclusions. Since the original paper was completed in December 1999 just before the Portuguese Presidency began, a Postscript has been added reflecting on some elements of the development of the European Union's policy in the employment-related field, especially those stimulated by what the author terms the emerging 'Lisbon Paradigm'.

2. THE KNOWLEDGE SOCIETY AND OTHER METAPHORS

A word of warning is called for. Any brief statement of the main elements which might be said to comprise the knowledge-based economy will almost inevitably move back and forth across the boundaries between description, analysis, prescription and speculation. This is certainly, in part, because the thesis is still being explored on its own terms and the data to test its applicability are hard to acquire. It is also because its potential implications are so important that they need to be examined within whatever framework of analysis can be articulated now and with the data available for scrutiny now.

Moreover, as demography, technology and globalisation generate their various imperatives, the policy community produces and reproduces initiatives which aim to counter the negative and enhance the positive potential effects of change. One new vision follows on from another and the basic ingredients of policy are recombined and then further 'refreshed' (as those who view the world through a web browser might have it) even within the attention spans of the policy community and the media, and during the life cycles of the policy products the one seeks to promote to the other.

The speed with which we have moved from the 'micro-electronics revolution', through 'information society' and 'learning society' to the 'knowledge society' is a case in point. These in turn are being increasingly overlaid by ideas which draw out the deeper implications of 'networking' and 'weightlessness' with which they are also now associated (see, for example, Castells, 1996; Coyle, 1997; and Kelly, 1998). Nouns and adjectives sometimes seem carelessly brought together, drawn out of a bag containing 'organisation', 'network', 'economy', 'society', 'age', 'information', 'learning', 'knowledge', etc. In another bag devoted to policy responses we find 'flexibility', 'skill', 'knowledge', 'competence', 'employablity', 'capability', 'enterprise', 'entrepreneurship', 'high-performance organisation', etc. Verbs also have their place in the lexicon of scenario makers: the economy is to be knowledge *based* or knowledge *driven*. And when all this is combined with the multitude of interpretations of 'globalisation', whether as a natural phenomenon or a policy choice by rich nations (Lindley, 1997a), we have a potent mixture for both confusion and glibness.

The visions switch their emphasis in rather subtle ways between acting as metaphors for a set of phenomena to which we must pay attention or to which we must adjust and conveying the nature of the adjustment being advocated. They vary, too, in the extent to which they deal with the 'economic' and the 'social'. Here, the *information society* is taken to refer principally to the enormous proliferation of information powered by the exploitation of micro-electronics and the first awakenings to its potential social as well as economic implications. The *learning society*, on the other hand, contains within it an embryonic design for modern living. This is heavily predicated on both the growing integration of information and communication technologies and fears of what globalisation is doing to European competitiveness. But it embodies flashes of positive thinking about the potential for widening as well as deepening the involvement of people in learning for life as well as for labour, throughout life as well as during the early years.

As for the *knowledge society*, what distinguishes it from the *learning society* is the view it takes of long-run structural change in the economy. The vision is that the production, dissemination and use of knowledge will take on a far more prominent role as a source of wealth creation and exploitation. There is much greater scope for codifying knowledge, abstracting it from its context, and this makes it potentially more accessible and marketable. At the same time individuals and organisations are likely to differentiate their performance from that of others by how they handle knowledge. An important aspect of this is the relationship between 'explicit' knowledge and 'tacit' knowledge. Organisations that are able to adopt ways of working that encourage the identification and sharing of key elements of knowledge which have hitherto been only tacit amongst their employees, sub-contractors, etc. are likely to be more effective. Thus, a radical diversification in the location of knowledge production within the sectoral and organisational structure of the economy is envisaged (Boisot, 1999). Many more organisations will quite genuinely be engaged in producing and disseminating as well as using knowledge.

The *learning society* and *knowledge society* scenarios have complementary implications for individuals as they progress through life and for the organisations with which they are involved. The former emphasises that learning should be a continuing activity. People should spend more time on it and the opportunities provided should be of a high quality, suited in content and delivery to the needs of different groups of learners. The latter points to the importance of recognising knowledge as an asset, the nature and location of which need to be carefully monitored and developed, along with the conditions that govern access to it. This is all the more crucial in periods of radical adjustment when an organisation's knowledge base can easily be damaged through poorly designed changes implemented quickly. Many of the failed attempts at business process re-engineering, for example, can be attributed to lack of attention to the relationship between human-resource and knowledge issues.

Up to about the early 1990s, the industrial world seemed to be moving in a direction where flexibility, shortening of contracts, narrowing of commitments and the replacement of principle by pragmatism would become the accepted way of conducting economic and, even, social life. Commitments to tackling social exclusion at the same time as these moves were afoot were evidently wearing thin. But, as the correction of past mistakes of macroeconomic policy began to bear fruit and, at least, some labour market rigidities were being addressed (see Section 3), the vision being explored started to alter in character and became what is now called the knowledge economy/society.

It is not then enough for people and organisations to embrace a 'learning culture'. New forms of management, participation, collaboration and contractual relationship need to be explored in order for them to prosper in a more knowledge-intensive environment. Several authors have argued that this will involve re-casting present approaches to human resources at various levels in the economy and society in order to get the right fit between the market for knowledge *per se* and the markets for work and learning.

Consideration of the knowledge society inherently blurs several boundaries which have hitherto governed our thinking, especially those between 'the economic and the social', 'the market and the organisation', 'competition and collaboration', 'companies and communities', 'consumption and investment' and 'work, employment, leisure and learning'. Yet this seems essentially to reinforce what has already been going on during the last decade and, in some cases, well before that. What is more significant is that the ground is shifting in several adjoining areas *together* and producing a *cumulative* effect upon the overall landscape of change. The policy and scientific communities will need to strike a balance between, on the one hand, communicating the importance of potential change so as to encourage adaptive responses and, on the other hand, providing some continuity of thinking so that people do not become disorientated.

3. LABOUR MARKETS AND THE KNOWLEDGE ECONOMY

The chapter by Boyer deals with the macroeconomic policy record and institutional frameworks for reconciling different economic and social objectives. Here, very brief reference in Section 3.1 will be made to the wider labour market context within which the development of the knowledge-based economy might occur. Section 3.2 then considers the main employment trends and projections with a particular focus on the US evidence because of the more advanced stage of transition apparent in that economy. Skill shortages and skills gaps are also noted in general terms but it is beyond the scope of this contribution to review the detailed evidence of the extent of these in the US and Europe. At the same time, as will be shown, one feature of the knowledge-based economy is that it lends a different perspective to looking at such phenomena and it is on this that the focus dwells.

More qualitative evidence is then summarised in Section 3.3 and the ideas of the knowledge-based economy explored in the light of the key features of occupational change which have been identified. In Section 3.4, the discussion moves from the occupational level to the organisational strategies that affect the structure of jobs and then to the institutional and policy context with which the section began. Finally, in Section 3.5, two particular issues are taken up: the role of professional highly qualified people in affecting change and the future place of the universities in the knowledge-based economy.

3.1. Job creation records, flexibility and employment systems

There are four main reasons why the record of the US is of particular relevance to the EU employment debate: (i) its performance in terms of output, employment generation, unemployment and average per capita income is far superior to that of the EU and, indeed, Japan, (ii) moreover, it has *increased* its lead in terms of growth and productivity, especially during the last decade, so its current position is not simply due to some legacy of the past, (iii) it has consistently adopted very different approaches to certain areas of policy compared with those of the EU, notably in the labour market, and (iv) the US exemplifies a high commitment to innovation and is much further on towards embracing notions of the knowledge-based economy.

About twenty-five years ago, the 'EU' (equivalent to the present member states) employment rate was slightly above and its unemployment rate substantially below the corresponding rates for the US. Since then the EU employment rate has fallen from 64 to 61 per cent while the US rate has risen from about 62 to 74 per cent. This, difference, in itself, need not be a problem since one of the best ways of benefiting from economic growth may be to enjoy more leisure and spend time on other non-market activities. A small fall in the employment rate could be quite compatible with growing well-being. However, among those who are recorded as members of the EU labour force, the unemployment rate has risen from 4 per cent to about 10 per cent and, on the face of it, that *would* seem to be a problem. In contrast, whereas the US unemployment rate was double that of the EU twenty-five years ago, it is now half the EU rate.

Nonetheless, there are other features of performance that need to be taken into account before we conclude that the overall employment situation in the US is one which Europe would really do well to aim for.

(a) The quantity of employment in terms of total hours worked per year, rather than simply people employed. Allowing for differences in working hours would actually increase the employment creation record of the US relative to that of the EU by about 4 per cent.

(b) The quality of jobs in terms of:
   - the pay and other conditions of employment,
   - the skills deployed,
   - the opportunities for advancement through experience and continuing, development.

(c) The duration of stay in poor quality jobs, allowing for recurrent spells in such jobs.

(d) The duration of stay in unemployment, allowing for recurrent spells of unemployment.

(e) Participation in measures to enhance short-term and long-term employability (assistance with job search, work experience, training, post-compulsory education).

(f) Wider socioeconomic conditions which may be linked, at least partly, with the labour market strategy of the US relative to that of the EU:
   - Those which seem to be fairly direct consequences of the strategy and are likely to carry over to the EU, even allowing for differences in underlying conditions and the specifics of implementation.
   - Those which are associated with the US employment situation but are more likely to stem from other features of its economy and society and would probably not carry over to the EU.

Essentially, the main conclusion from recent studies of the US experience which try to explore the above aspects is that there is a substantial problem with job quality, inequity *and* social exclusion. In other words the vision of dynamic job creation in which there are many poor jobs but that these serve as the initial rungs on the ladder of upward occupational mobility does not quite ring true: higher inequality within the population or among households at a *point in time* does, after all, mean higher inequality as measured over *life cycles*.

There are many comparisons of economic performance among the industrialised economies. The OECD's (1994) *Job Study* is especially important because of the extent of its analysis, the fact that it concluded by placing great emphasis on differences in labour market regulation as an explanation for the relatively poor performance of Europe, and the OECD's subsequent follow-up of progress in adopting its recommendations. This is not to say that there has been exactly complete agreement that the OECD has assessed appropriately the evidence or that its preoccupation with promoting 'labour market flexibility' has been well judged. The

evidence which is accumulating would now seem to suggest that the importance of de-regulation has been overstated (Grubb and Wells, 1993;
Lindley, 1997b; OECD, 1994 and 1999; Nickell and Layard, 1998).

A broad conclusion would be that the main reason why unemployment
is now lower in the US relative to Europe is because compensation for lack
of work via unemployment insurance ceases much earlier in the US. This
represents a social policy choice and has less to do with the relative degrees
of labour market regulation.

A further conclusion is that what is substantially affected by regulation
is not the *level* of employment in relation to the population of working age
but the *pattern* of its distribution among different forms of work (degrees
of security – 'permanent', open-ended, fixed term, temporary, casual –
employee and self-employed; full time, part time, overtime, shift work,
Sunday work and other working hours arrangements) and among different
socio-demographic groups. So labour market regulation and the terms governing social security benefits both primarily affect equity not efficiency.

3.2   Occupational employment trends

Differences in the performance of European and US innovation systems,
the growth of ICT-based industries and ICT-related services such as
e-commerce are dealt with in the chapter by Soete. The aims of the present
paper are to examine the overall patterns of employment in quantitative
and qualitative terms from the perspective of ideas about the knowledge-
based economy. Since the US government provides much more extensive
information and analyses of potential changes in employment and the US
is well ahead in the kinds of innovation that are leading to the knowledge-
based economy, it should be especially useful to examine the main features
of the latest projections released in 1997 (Table 4.1).

In broad terms the fastest growth is expected to arise in managerial, professional and technical occupations, all of which are associated with high
levels of education and training, and among so-called 'service workers', a
rather hybrid category comprising medium-to-low levels of skill, depending on the service activity in which they are engaged. More detailed projections show that there is great diversity among the fastest growing
occupations as regards their industrial context and education and training
requirements. Certainly IT-related occupations are in the top three positions: 'database administrators, computer support specialists and other
computer scientists', 'computer engineers' and 'systems analysts', all of
which categories are projected to double their employment in the decade to
2006. But the picture is not simply one of burgeoning growth in occupations directly related to the IT sector or to specialist IT occupations

Table 4.1 Employment and total job openings 1996–2006 by education and training category, United States

Education and training category	Employment						Total job openings due to growth and net job openings	
	Thousands		Per cent distribution		Change		Thousands	Per cent distribution
	1996	2006	1996	2006	Thousands	Per cent		
Total, all occupations	132,353	150,927	100.0	100.0	18,574	14.0	50,563	100.0
Bachelor's degree or higher	28,885	35,207	21.8	23.3	6,321	21.9	12,296	24.3
First professional degree	1,707	2,015	1.3	1.3	308	18.0	582	1.2
Doctoral degree	1,016	1,209	0.8	0.8	193	19.0	460	0.9
Master's degree	1,371	1,577	1.0	1.0	206	15.0	430	0.9
Work experience plus bachelor's or higher degree	8,971	10,568	6.8	7.0	1,597	17.8	3,481	6.9
Bachelor's degree	15,821	19,838	12.0	13.1	4,017	25.4	7,343	14.5
Post-secondary education or training below the bachelor's	12,213	13,725	9.2	9.1	1,513	12.4	3,943	7.8
Associate's degree	4,122	5,036	3.1	3.3	915	22.2	1,614	3.2
Post-secondary vocational raining	8,091	8,689	6.1	5.8	598	7.4	2,329	4.6
On-the-job training or experience	91,256	101,966	68.9	67.6	10,740	13.1	34,323	67.9
Work experience in a related occupation	9,966	11,177	7.5	7.4	1,211	12.2	3,285	6.5
Long-term on-the-job training	12,373	13,497	9.3	8.9	1,125	9.1	3,466	6.9
Moderate-term on-the-job training	16,792	18,260	12.7	12.1	1,468	8.7	5,628	11.1
Short-term on-the-job training	52,125	59,062	39.4	39.1	6,937	13.3	21,944	43.4

Source: Silvestri (1997).

elsewhere in the economy. Indeed, perhaps the most striking findings are that over half of the top twenty growing occupations are associated with the health sector and over half require education and training which is significantly *below* that of a bachelor's degree. This is further reinforced by examining the occupations which account for the largest absolute increases in employment. The diversity is then even greater and more than two-thirds require less than a bachelor's degree; over half actually only require short-term on-the-job training.

In fact, looking at the overall assessment, we find that a third of employment growth is in occupations that require a bachelor's degree or higher and a third require only short-term on-the-job training. Of course, great care is needed in the interpretation of projections of the growth of different occupations. Aside from the uncertainties attached to such exercises, the projections do not indicate the job openings that are likely to arise since they do not allow for mobility between occupations and retirements. When these are taken into account, the job openings for those with bachelor or higher degrees account for almost a quarter of the total and those needing only short-term on-the-job training account for over 40 per cent.

Concern over skill shortages has inevitably arisen as the US has experienced roughly eight years of expansion. Most prominent in the media have been claims of shortages in the IT area but these have also been accompanied by reports of shortages of construction labourers and craftsmen, registered nurses, and teachers. Veneri (1999) concludes that dramatic stories of shortages, high earnings potential and an over-heated hiring climate for information technology workers do not really square with the facts. The evidence is more in keeping with a conclusion that *stresses the similarities of their situation with that of professional specialists in general.*

Of course, this does not mean that significant shortages may not arise in the future, whether in the US or the EU (though the supply-side behaviour of these countries differs greatly). What is worth noting, however, is the similarity of the diversity of occupations involved when identifying the sources of employment growth in the US and those identified in a wide range of more *ad hoc* studies for the EU. The need for much more systematic European assessments of future scenarios will be discussed in Section 5. However, selections of findings for Sweden taken direct from the official web site and for certain other member states taken from a recent *ad hoc* projection exercise are presented in Tables 4.2 and 4.3. They show the importance of the growth of management, professional and intermediate occupations, but also the hybrid 'services' mentioned above in connection with the US. They also display considerable variation in the absolute and relative growths of occupations across countries.

Taking the evidence reviewed so far, one particular sectoral perspective

Table 4.2   Potential bottleneck occupations in Sweden up to 2010

Industry
- Graduate engineer (most specialities)
- Graduate technician (most specialities)
- Chemist
- Physicist
- Mathematician
- Welder (trained specialist)
- Toolmaker
- Engineering worker (all-round)
- NC/CNC operator
- Printing worker (new technology)
- Electrical fitter
- Plater

Construction
- Architect
- Structural technician/engineer
- Bricklayer
- Concrete worker
- Joiner
- Construction worker
- Roof fitter
- Floor-layer
- Plumbing and heating installation worker
- Craftsman electrician
- Painter

Private services
- Tele-salesperson
- Advertising salesperson
- Electronics salesperson
- Corporate salesperson
- Logistician
- Logistics engineer
- Aerospace technician

- Ship's officer
- Machine officer
- Bus driver
- Economist (4-year post-secondary programme)
- Cook
- Construction plant operator
- IT specialist (all specialities)
- IT engineer
- Web designer
- Public relations officer
- Advertising/multimedia
- Auditor
- Lawyer
- Interviewer
- Organisational consultant

Public services
- Upper secondary school teacher (general subjects)
- Vocational teacher (IT)
- Vocational teacher
- Special teacher
- University/college teacher
- Doctor
- Nurse (with further training, various specialities)
- Nurse
- Assistant nurse
- Psychologist
- Dentist
- Dental hygienist
- Pharmacist
- Dispensing chemist
- Police officer
- Fire safety engineer
- Vocational counsellor

Source: Swedish National Labour Market Board (1999).

Table 4.3   Selected occupational projections for the EU, 1996–2005 (% change)

Occupation	Germany	Ireland[b]	Netherlands[b]	Spain	United Kingdom[b]	EU9[a]
Managers	–	23	14	20	12	14
Professionals	23	28	9	11	7	5
Technicians	20	41	10	95	7	20
Clerks	−5	6	−12	−31	1	−5
Service workers	5	13	6	17	18	13
Agricultural workers	−7	−23	−47	−36	−11	−15
Craft workers	−10	10	4	11	−21	−1
Plant workers	3	10	9	28	3	4
Elementary	−15	12	22	−19	−7	−4
Total	4	13	6	12	4	5

Notes: [a] Excludes Austria, Finland, Greece, Luxembourg, Portugal and Sweden, mainly for reasons of the availability of suitable occupational data or the country coverage of the underlying multicountry, multisectoral model, E3ME (developed by Cambridge Econometrics) at the time of making the initial sectoral projections.
[b] Regular projections for 5–10 years ahead are made annually for these three countries by the ESRI, ROA and IER, respectively. They give more detail at the occupational level but their conclusions broadly reflect those given in the text and in the summary of a wide range of studies given in Table 4.4. See Sexton, *et al.* (1998), de Grip and Heijke (1998), and Wilson (ed.) (1998) for presentations of the results and methodologies used for the three countries.

Source: Projections conducted by the Institute for Employment Research, University of Warwick, and Cambridge Econometrics, in consultation with several other organisations. See Lindley (1999) for a description of the overall study. Note that such projections are highly sensitive to a large range of assumptions about both the world economic environment and behavioural responses in the individual economies. They are also especially sensitive to the classification of occupations used for the European Labour Force Survey, which changed in 1992. Some of the large relative differences in national trends for the same occupational category will be due to the inevitable presence of definitional disparities and transitional effects due to the change of classification. See also CEC (1998). The projections should thus be regarded as being only very broadly indicative of possible future changes. Further details will be accessible direct from the IER web site in due course (http://www.warwick.ac.uk/ier).

consistently emerges. Strategies which increase the knowledge-intensity of services and not just the skills of those supplying services in their *existing* forms will be especially important. The potential demand for more education, training and continuing professional development may neither materialise in the first place nor be satisfied adequately without major reform. What should the balance between state and private funding be? What options are most promising for the role of government in modernising and regulating markets for such services so as to increase demand, quality of service and the quality of jobs? Health and personal care are also key areas where rising demand through population ageing sets the scene for opportunities to implement more knowledge-intensive, job-creating, yet inclusive approaches to personal services.

3.3 Qualitative dimensions and the concept of practice

The examination of likely employment trends in the previous sub-section needs to be given a more qualitative perspective. Reviewing occupational change in industrialised countries over the last two decades or so, several features stand out in virtually all such economies. These are summarised in Table 4.4. They may also be taken together with another set of quite common features of the skills debate in different countries that relate to the emergence of concern about so-called *generic* skills; these have also been termed 'basic', 'core', 'key' or 'transferable' skills but are now usually seen to relate to:

- communication – literacy,
- application of number – numeracy,
- problem solving,
- working with others,
- improving own learning and performance,
- knowledge of information technology – computer literacy.

There is some controversy over defining these, over how best they might be acquired and over what value the labour market actually places upon them in practice (as opposed to the significance attached to them by employers when responding to surveys and in case study interviews) (Green, 1999). The evidence is rather weak on the extent of the growth in importance of generic skills relative to job-specific and occupation-specific skills (those linked, respectively, to the particular situation of an employer or to a particular occupation carried out in ways which are more or less common to many employment situations). On the other hand, reliable methods of monitoring their significance have yet to be devised and incorporated into suitable survey instruments.

For those jobs that do survive change and those that emerge from it in quite new forms, there is a broadening of the context in which people work, an increasing range of options as to how to do the job as it is, frequent re-assessment (though often implicit rather than formal and explicit) of its relationship with other jobs in the organisation as new tasks appear on the

Table 4.4   Qualitative aspects of occupational change

Occupations concerned with planning, monitoring and control – notably, managers and supervisors
Changes in job content which will accompany the further expansion of management jobs will reflect the wider range of competence expected of managers and the need to take responsibility for a more complex process. The latter will use more capital and fewer people but with personnel engaged on more demanding tasks and involving greater autonomy. The requirement for broader expertise will affect general managers and specialists alike with the former acquiring greater technical knowledge and the latter more business skills. Responsibility for a greater mix of capital equipment and skilled employees will also characterise the supervisory occupations where more emphasis will be laid upon communication with and the motivation of employees.

The relationship between higher and intermediate professions
This issue concerns the relationships both between long-established occupations (doctor/nurse, scientist/laboratory technician, etc.) and between relatively new occupations (software engineer/programmer, design engineer/technician/ draughtsperson, manager/data processing specialist). In many of these linked occupations, changes in product demand, labour market conditions and vocational education are creating situations in which significant adjustments have taken place in some countries. Relative salary costs and shortages of certain higher-level skills combined with the development of technological aids to decision making indicate the potential for further changes (e.g. in health care) subject to institutional restraints.

In addition, in certain areas of the economy there seems to be scope for the emergence of new higher and intermediate professional groups. For example, the development of tourism and leisure activities need not be associated with a continuation of temporary, sometimes seasonal, low-skilled and low-paid employment. Aiming at higher quality and a wider range of services generates a demand for more professionally qualified entrepreneurs, managers and local administrators, supported by a cadre of skilled personnel. The latter may work behind the scenes or be in direct contact with the customer (e.g., in providing tourist/travel information, instruction in recreational pursuits, supervision of customer relations in hotel, catering and recreational establishments). The speed with which such service industries move towards the 'high value added' end of their product market clearly determines the rate of growth of these more-skilled occupations. It is also, however, determined by the rate at which appropriate new professional roles are identified and the relevant training is provided.

Other non-manual occupations
This group is dominated by clerical, secretarial and sales occupations. Some of these personnel are directly involved with customers in areas where attempts are being made to improve the quality of customer service. There the job content is

Table 4.4   (cont.)

---

placing greater stress on the combination of a higher level of product knowledge, inter-personal communication skills, keyboard skills and software knowledge needed to use the financial/sales information system, and wider commercial awareness.

In other respects, though, there seems to be some uncertainty about the net impact of organisational and technological change upon these occupations. Many routine clerical tasks and the supervisory roles attached to them have been abolished with the widespread establishment and use of machine-readable data bases. The scope for more complex analysis and presentation of data and the ability to create higher-quality documentation is likely to increase the demand for certain clerical/secretarial staff. These will have higher levels of literacy and numeracy, greater knowledge of the business and its information system, and skills in the use of software for word processing, statistical analysis and graphics.

Quality considerations mentioned above in relation to customer service combine with efficiency arguments to create a demand for more flexible personnel. These should be able to switch from counter service to liaising with suppliers to carrying out supporting clerical and secretarial tasks. This is an emerging form of multiskilled and multifunctional office-based occupation which has received less attention than does its manual craft counterpart (see below).

Skilled production occupations

The fate of skilled craftsmen has been crucial in the implementation of new technology. The possible consequences are: (a) abolition of the job; (b) de-skilling to a large degree (e.g., to a machine-minding or materials handling function); (c) re-skilling where previous knowledge is transferred to a new context requiring some re-training (e.g., transfer to maintenance from production); (d) multiskilling where, typically, the need for a wider craft-based competence involves the acquisition of complementary skills in electronics; and (e) up-grading to the status of technician/programmer/complex keyboard operator which exploits previous knowledge but involves substantial re-training or recruitment of new qualified staff.

The development of the multiskilled, multifunctional, worker has been highlighted in much case-study evidence. A worker who deploys, for example, mechanical and electronics skills and deals with production, regulation of equipment and minor repairs would fall into this category. Similar attention is paid to the specialised maintenance and repair functions employing highly skilled craftsmen. However, the introduction of new technology generally creates far fewer such jobs than it destroys or de-skills traditional craft jobs engaged in production. The extent of up-grading has evidently become the litmus test of enlightened management. But the evidence on the relative costs of alternative strategies as opposed to their feasibility in terms of the technical and training requirements is extremely limited. The emerging division of labour between skilled manual workers and more highly qualified technical personnel at the point

Table 4.4   (cont.)

---

of technological transition is insufficiently monitored and understood at national level to allow for a satisfactory explanation of international differences.

Management control versus technological necessity as determinants of job content and occupational structure
The above point implies that it would also be premature to generalise on the much debated issue of how far employers seek out technologies and choose modes of implementation which increase their control over labour as the principal aim. Nonetheless, the introduction of information technology (in its widest sense) does present a particularly important opportunity for examining this issue. Its flexibility would seem to offer the potential for a much greater variation of economically acceptable management choices than with other major technologies.

Source: Author's assessment of a wide range of *ad hoc* qualitative as well as quantitative studies

boundaries between them, and periodic reviews of the activities undertaken by the organisation as opposed to those that are or could be undertaken by suppliers and customers.

The implication of the above is that workers increasingly do have a notion of their 'practice' derived from reflecting on what they do and why they do it that way rather than another way. Reflection may, moreover, bring out the *tacit* knowledge that the individual employs in acting in a particular situation but which is inherently more difficult to communicate than the explicit knowledge or skills that form the focus of education or training. For that reason it may be the factor that differentiates the performance of one individual from another when access to other forms of learning is otherwise the same. And at the organisational level, if employees can actually *share* tacit knowledge, it may be a key factor in determining overall performance and success *vis-à-vis* competitors.

If the individual is to reflect on her or his practice and share those reflections with other workers, it is a relatively short step to reflection on the practices of those around them. Not all tacit knowledge may be seen to have equal value just because it is tacit. The opportunity to learn from others is tempered by the threat of discovering that one's own practice is rather inferior. Thus the kind of cooperation and trust required to identify and share what would otherwise remain tacit knowledge cannot be assumed to be present automatically. Nor can it be assumed that workers who may be willing to share with each other in very informal ways would wish to share with the organisation. It is here where the attention paid to the nature of knowledge and skills in relation to occupations comes together with that paid to the idea of the high-performance organisation and to the conditions likely to give rise to their creation and flourishing in the modern economy.

3.4   Organisational styles, social partnership and the high-performance organisation

At the heart of the issue is the tension between encouraging individuals and work groups to reflect on *their* practice and refine the knowledge they have, communicating it to others in the interests of the organisation as a whole and *a reluctance of the organisation itself to reciprocate.* The deal needs to be different in order for the organisation to adapt to changing external conditions and to initiate from within change which will require other organisations to respond.

A *learning organisation* is one in which people are able not only to update continually their knowledge and skills but also to learn from the mistakes they will nonetheless make individually and collectively, however competent they are. The *knowledge-based organisation* is located in or across sectors of the economy that trade intensively in the creation and exploitation of knowledge. A crucial competitive advantage will be how it handles its own knowledge of what it is doing and the intangible assets intimately associated with its main activities, especially its intellectual capital (Stewart, 1997). Its other assets, such as possession of highly valued physical assets, access to cheap energy sources, favourable spatial location with respect to suppliers and customers and even current market position will be of much less importance. In these circumstances the knowledge-based organisation must be a particularly effective learning organisation, but the central human resource management task is to *manage knowledge.* Indeed, the research focus on the learning organisation has been gradually replaced from the mid 1990s by that on knowledge management (Scarborough, Swan and Preston, 1999).

The importance of generic skills seems to lie in the capacity to cope with and contribute to the change process. Job-specific and occupation-specific skills do not adapt easily without broader knowledge, skills and attitudes to change. At this point, the consideration of occupational change and the widespread importance attached to generic skills meets the discussion of organisational styles that will promote a knowledge-based economy. The scenario is one in which, over increasing areas of the occupational map, people are being asked to take more responsibility for their work situations: for getting the knowledge and skills to do the job in the first place and updating them appropriately, ensuring they have the information they need to carry out their jobs, responding creatively to changing requirements, and taking the initiative when they see potential for developments which will benefit the organisation. The organisation must, for its part, facilitate this process, i.e. 'empower' the individual and work group to play this role.

Notions of the high-performance organisation began to emerge in the employment policy debate partly as an antidote to simplistic ideas of what to expect from labour market de-regulation and the preoccupation with 'unburdening management' by removing constraints on their prerogative to manage. Much less attention was given to the idea that there were shortages of good managers and good organisational designs and that this might have equal or greater significance for the performance of the labour market and the economy at large. The contribution of the European Commission and some member states in trying to enrich the social dialogue process by seeking shared understanding of what forms of organisation seem to work has its counterparts in the US, although under very different institutional conditions. The Commission on the Future of Worker–Management Relations, established by the Clinton Administration in 1993 particularly sought to identify the characteristics that mark out the high performers. One of the Commission's members (Marshall, 1994, 1998) has summarised these and they are reproduced in Table 4.5. More recently, Gray, Myers and Myers (1999, p. 29) report the results of a survey which point to a 'clear trend . . . in which firms are progressing towards a new collective bargaining paradigm' involving some sort of 'partnering agreement'. The 'Compact for Change' issued by the labour-management Collective Bargaining Forum in 1991 provides the structure for the survey analysis and is shown in Table 4.6. Such ideas would not be out of keeping with European social dialogue themes promoted during the 1990s and are clearly in tune with the organisational characteristics sought in the knowledge-based economy scenario.

3.5   The knowledge-based economy: the roles of professionals and universities

Harnessing the expertise of the highly qualified – a new professional ethic?
There are a number of cross-cutting themes which emerge from discussions focusing narrowly on the knowledge-based economy and those which look at its implications for society. The importance of the so-called 'professional project' has clearly increased and will do so further, but the principles determining how it should be handled are still very much un-debated. The work of Schön (1983) and collaborators, for example, focused on the nature of professional learning ('knowing in practice' and 'reflection in action') and the institutional and market contexts in which it takes place, but has not

Table 4.5   Characteristics of high-performance organisations

High-performance organisation	Mass production system
1. Quality driven and therefore establishes closer and more cooperative relations with customers and suppliers	Producer driven and plays suppliers off against each other through price competition
2. Lean management structures that promote horizontal cooperation and participative management styles: establishes mutual trust and respect between managers and workers, and decentralises decisions to the workplace.	Hierarchical, segmented 'Tayloristic' approach to management.
3. Stresses internal flexibility to adjust quickly to changing technology and markets. Stability sought through quality, productivity and flexibility.	Stability sought through rules, regulations and contractual relationships.
4. High priority given to positive incentive systems, relating rewards to desired outcomes. Important because the efficient use of leading-edge technology gives workers considerable discretion.	Stresses negative incentives (punishment and layoffs) or even perverse incentives which make it more difficult to achieve desired outcomes: e.g., when workers believe improving productivity incentives will cost them their jobs.
Positive incentives include: bonus schemes, internal cohesion, fairness and equity; job security; and participation in decision making.	Hierarchical arrangements, fragmented work, and adversarial relations discourage the cooperation required for high levels of quality and productivity.
5. Develops and uses leading-edge technology through constant improvement on the job and by adapting advanced technologies produced elsewhere. Realises that standardised technologies mean competing mainly on wages.	
6. Pays heavy attention to education and training for all workers. Higher-order thinking skills are required for high performance and the development and use of leading-edge technologies.	Stresses education and training for management and technical elites.

Source: Derived from Marshall (1994).

Table 4.6   A compact for change

---

1. The parties should jointly work to increase productivity and enhance the quality of products in order to ensure employees of long-term security and a rising standard of living.
2. Management should reflect the continuing improvements in productivity and the quality of products in its decisions regarding worker compensation, the organisational structure, pricing and investment.
3. Unions and employers should jointly develop the leadership and technical skills of their workers.
4. Unions and employers should jointly develop ways to promote teamwork and employee involvement in developing and administering personnel policies and in strategy decision making to achieve organisational goals.
5. The employer and union must commit to open and early sharing of information relevant to corporate strategies and the relationship between the parties.
6. The employer and union should share their views and agree on how employee representation will be determined at new facilities.
7. Permanent separation of workers will be an action of last resort.
8. Both unions and the employer should be jointly committed to a work environment in which disputes are resolved in an amicable manner, without resort to strikes, lockouts or hiring replacements.
9. Both parties should be committed to worker participation at all levels of decision-making in order to provide continuous improvements in products, services, safety, employment security and productivity.
10. It is essential that employees have input in the design and application of new technology and in the planning and development of any new system for the allocation of tasks.

Source: Gray, Myers and Myers (1999), taken direct from *Labor-Management Commitments: Compact for Change* (US Department of Labor, 1991).

considered how different interpretations of the role of, in Reich's (1991) terms, 'symbolic analysts', would have quite different consequences for the relationship between economy and society.

In particular, the potential enrichment of occupational roles of those supporting the work of professional occupations comes up against the potential 'hoarding' behaviour of professionals when facing risks attaching to their own positions and the facility offered by ICT to substitute their own work for that of others, even if this represents a denial of the basic principle of comparative advantage.

In addition to the possibility of professional hoarding at the individual level, there is the powerful instinct in a number of occupational areas in favour of professional 'solidarity'. This consists of a *reluctance to differentiate fundamental roles and levels of competence within the profession*, as well as defending the profession's position from competition from the intermediate or para-professional level. This, inevitably, delays or blocks the development of modular structures of education and training, as well as rigidifying the occupational structures to which they relate. Overlapping career paths which would facilitate upward mobility are thus poorly developed in a regime characterised by professional hoarding and solidarity.

In terms of the economy, the consequences of the market power already held by certain groups can be a misallocation of the highly qualified, leading to bottlenecks at higher and lower levels of professional work, which limits the scope of job generation further down the occupational hierarchy. This leads to having the worst of both worlds.

In the dynamic context of the knowledge-based economy it seems that a new professional ethic may be required to match that of 'work sharing' hitherto, often in the context of agreements between the social partners. This ethic is that of 'knowledge sharing' in which an obligation is placed upon professional groups to seek out ways of structuring work so as to generate occupational profiles that reach across the gap between professional employment and skilled industrial or office-based work. In constructing these occupations, it is essential to avoid creating new types of barrier to entry which undermine the scope for achieving equal opportunities at the same time as benefits to business. It is not too far-fetched to think in terms of 'fair employment structures' and a forum or agency responsible for identifying problems of access to and limited development of different parts of the occupational structure.

Thus many of the possible paths on the new and evolving landscape cut across the vested interests of the growing numbers of highly qualified professionals. Yet governments will need to harness much greater commitment from professional groups to think creatively and in socially responsive ways about how to reap the benefits of the knowledge-based society. How we promote occupational enrichment and open employment structures rather than produce polarisation within closed structures will be an especially important issue.

At the same time, the relationship between employment in an occupation and the focus of 'formation' differs greatly between areas such as the law, medicine, education, management, engineering and science. In periods of growth, the work of science postgraduates can in effect remove the need for a substantial part of what would otherwise be more conventional intermediate-level employment, but there then arises a problem in situations where the knowledge production system, having relied upon students, loses its attractive power and has rather poor arrangements in place to replace them with more fully thought-out occupational roles.

The relevance of this to the knowledge-based economy is that an increasing number of activities will begin to look like one or more of the knowledge production contexts found in universities and *vice versa*. Indeed, at a more general level this has been one factor behind the creation of 'corporate' universities by certain companies. It is also relevant because of the increasing importance that is likely to be placed upon the role of the universities in knowledge-based economies.

Learning, labour and innovation

In the knowledge-based economy, the learning, labour and innovation systems will be linked much more intimately. The first of these will not simply provide highly skilled labour, basic research and strategic research for the innovation system to incorporate into products and processes. This linear model has long been out-dated and many of the leading European universities have been involved with industry for many years (Gibbon *et al.*, 1994). The key characteristics of this relationship at its best has been *interaction*. Moreover, this needs to be the norm for the future rather than a characteristic displayed by particularly innovative universities. Universities, or their equivalent, will be transformed by far more 'engagement' with

- other major producers and users of knowledge outside the university and other parts of the publicly funded research system;
- a variety of intermediaries concerned with more intensive knowledge transfer activities for and with small and medium-sized enterprises (SMEs);
- continuing professional development building on (inter)national reputation, regional and local association as well as alumni relationships, in which 'graduation' is merely the start of a process of deepening and broadening the knowledge base and higher-level generic and occupationally specific competencies sought by the individual, with varying degrees of involvement of employers;
- regional economic development strategies in which they will play an active part not merely in helping to implement but in helping to construct;
- and local communities.

The importance of the transformation of universities from exclusive producers of knowledge and educators of the highly qualified to partners in complex networks which go far beyond the pedagogical and scientific boundaries of their traditional roles should not be underestimated as a major force in the development of the knowledge-based economy and society. Great changes in culture, management and structure of these institutions will be needed if the societies within which they are embedded are to benefit fully from their potential. Moreover, the contrast between the allegedly aloof world-class universities and their lower-class cousins who participate regionally and locally to make up for lack of academic excellence will be seen for the myth it really is. The best universities already make enormous direct as well as indirect contributions to their regions. Those institutions which wish to emulate them will need to take this role on board as much as the conventional academic profile they naturally aim to pursue.

Essentially, universities have an opportunity to become major animators and participants in networks of *practice* as well as networks of knowledge production and dissemination. Moreover, engagement with the world of practice will be required because the professional development role will only be sustainable if the university staff include a strong contingent of those who are experienced *in professional practice* as well as having the necessary academic track record. The meaning of this will differ from discipline to discipline and will inevitably have implications for the balance of power between disciplines and, indeed, between them and multi- or inter-disciplinary endeavours.

An important issue for universities and other educational and training institutions will be the extent of their involvement in and the division of labour between them in the initial training and continuing development of *para-professional or intermediate occupational groups*. This is partly related to the fact that, as higher education expands, increasing proportions of graduates are finding not just jobs but careers in these areas of the occupational spectrum. This generates a supply-side impetus for changing the profiles of the existing occupations somewhat. In some areas this is reinforced from the demand side as both the knowledge base for their work and the range of functional roles they play are extended. There is then a potential overlap between the routes to the 'formation' of these new cadres and those of the professional groups with whom they interact. This raises questions about the appropriate institutional location of different parts of their education and training and the incorporation of work experience. It all places the structuring of the profession concerned on the agenda.

This also draws higher and further (non-higher, post-secondary) educational institutions deeper into the process of *professionalisation* of certain groups. There is great potential for collaboration between the two sets of institutions in delivering an increasingly diverse range of vocational higher education as new professions and para-professions form out of the evolving job structure. This would reinforce the kinds of collaboration which have developed between institutions of further and higher education in order to facilitate the greater access of those social groups who are still poorly represented in higher education.

4.   SOCIAL EXCLUSION AND THE KNOWLEDGE ECONOMY

This section deals with social exclusion only in relation to the labour market, skill acquisition, and economic development. The chapter by Esping-Anderson covers the sustainability of European welfare states overall, including a discussion of population ageing.

4.1   Skills and qualifications: relationships between objectives

Previous research and policy statements at the national and EU level have stressed the scope for achieving greater competitiveness through increasing skill and the overall enhancement of the qualifications and competencies embodied in the population. They have also identified the particularly important mechanisms by which industrial re-structuring can be hastened by paying more attention to the human capital needs of those sectors which are seeking product market positions in higher value-added segments.

Equally, where it is necessary to redeploy the labour force from industries with rather poor prospects of movement up the value-added hierarchy to those with better prospects, there is a potentially key role for vocational education and training in helping new industries to grow more rapidly than would otherwise be the case and in assisting members of the labour force to switch into the jobs being created.

However, the pursuit of faster growth through investing in the skills needed by developing industries and through minimising the drag on the economy from too slow a withdrawal from less promising sectors will not necessarily have much effect on social exclusion. Unemployment in Europe is too high to assume that the natural process of recovery and expansion in traded goods and services will draw back into productive employment those who are most marginalised. The queue is too long to rely upon this mechanism. Nor can it be assumed that much more competitive manufacturing and traded services allow the more labour-intensive activities, producing mainly non-traded goods and services, to expand through fiscal stimuli in the form of higher public spending and lower taxes.

4.2.   What is social exclusion?

Furthermore, the relationship between unemployment and social exclusion is a complicated one. Not all the unemployed are socially excluded, some are just unemployed. Some of the excluded are employed and whilst their jobs are not confined to the informal sector, their work is characterised by a degree of uncertainty about continuity and the pay and conditions received, which is reflected in the precariousness of their economic lives.

On the other hand, the socially excluded are not just the 'working poor', the long-term unemployed and their dependants. There are several ingredients to their situation. They lack citizenship in the political sense as well as in social and economic terms. They lack access to power through those able to represent their interest and improve their status. They lack access to resources in general because the public institutions are not at their disposal but are rather more concerned to contain or control the problems which the socially excluded 'create'. They have limited access to education and training beyond the stage of compulsory schooling. They lack access to networks which help to mobilise mutual support and share information and advice. They lack access to an effective financial and social 'safety net'. Overall, the socially excluded suffer from cumulative disadvantages that frustrate their attempts to escape and gradually reduce their motivation to try out new strategies.

If we consider the contours of socioeconomic exclusion, there are familiar associations between certain characteristics of individuals and groups and the extent to which they live at the margin of society. Certain combinations of such personal characteristics as gender, age, disability, ethnicity, nationality or citizenship status tend to be associated with forms of social exclusion. The same is true of family or household characteristics relating to family composition, personal attributes of other family members, position regarding income and wealth and spatial location. So, for example, those living in single-parent families, households with other members who are unemployed, less-developed rural areas, disadvantaged inner city areas and poor climactic environments are likely to be at risk of persistent marginalisation.

Whilst personal and family characteristics may be strong determinants of the probability of social exclusion, especially among young people, individual experience is also relevant. The extent of effective socialisation, education and training is of special importance: those provided with basic literacy, numeracy, social skills and the cognitive and manipulative skills required in reasonably promising occupational areas are less likely to experience a succession of precarious jobs. But, regardless of such preparation for working life, the effect on both the individual and prospective employers of a recurrence of spells of unemployment may be such as to discourage the former from trying to find work and predispose the latter against offering it. Finally, the extent to which a person has engaged in non-market activity will affect their labour market potential. In particular, the valuation of competence displayed outside the market place is typically lower than is justified, especially in the case of women seeking to re-enter the labour market after a period involving the care of children or an infirm relative.

4.3. How does social exclusion arise? – How is it fostered?

From the above, we might conclude that there are several ways of becoming socially excluded. It is possible to be born excluded, to slide into exclusion through the declining fortunes of one's family or one's own working life, or to be overtaken by exclusion triggered by political or economic change. In the last of these, political or legislative change may alter the status of particular groups (such as migrants or certain classes of nationals) or major reversals in the conditions governing a sector may dramatically affect the prospects for some of its workers, plunging the youngest and oldest into prolonged unemployment. Of course, cohesion of the local community may be sufficient to prevent the latter leading to exclusion; as noted earlier, unemployment is neither a necessary nor sufficient condition for exclusion.

Whilst social exclusion may arise out of an unfortunate combination of circumstances that propels certain people or groups to the margins of a society, there are structures that foster exclusion. Mention has already been made of social and political factors but there are policy stances which tend to create or fail to remove conditions that militate against the fortunes of the most vulnerable social groups. First, even with extensive public education, the tendency for the system to stratify itself has led to very poor preparation for working life for the least able in many countries. Moreover, there is a tendency for access to initial and continuing vocational training to be determined mainly by the possession of an employment contract. This means that in a period of high unemployment some members of the labour force will experience great difficulty in obtaining adequate training to enable them to get secure employment.

Second, segmentation and discrimination in the labour market serve to reduce the opportunities for some groups almost regardless of how capable they are of doing the jobs available. Third, the way in which pay and other conditions of employment are structured tends to generate marginal job categories even in sectors offering high-quality jobs. Limited job security, low pay and exclusion from the company pension scheme lead to poor pensions and the risk of impoverishment during retirement, when the opportunities for supplementing income through taking paid work decline markedly.

Finally, the experience of prolonged recessions and the persistence of high unemployment in Europe means that the labour market structures that foster social exclusion have had even more marked effects, particularly the growth of long-term unemployment and the creation of high levels of youth unemployment. These conditions are unfortunately highly conducive to reinforcing tendencies towards exclusion.

4.4   Social exclusion and skill

This brings us to the relationship between social exclusion and skill. The current situation in the EU presents a number of problems of diagnosis. The presence of a disproportionate number of unskilled and unqualified people among the unemployed has led some policy advisers to conclude that the remedy is to provide more educational and training opportunities for the unemployed. Data are frequently presented to show the much higher rates of unemployment and long-term unemployment among the least educated and trained (Figure 4.1), the longer cumulative lifetime spells of unemployment which they experience (see the estimates for men in Table 4.7), and their very low relative income (Figure 4.2).

However, this does not take sufficient account of high levels of unemployment and the extent of 'educational crowding out'. If the untrained are unemployed because more qualified people are moving down the job hierarchy in search of work, displacing the less qualified, this does not mean that training the most disadvantaged unemployed will substantially improve their employment prospects or their productivity in those jobs they are able to obtain. Their competence to do the jobs is not necessarily in question, it is their position in the queue which is enhanced at the expense of someone who is even less well qualified. In these circumstances, training the unemployed may simply concentrate further the incidence of unemployment upon those who do not receive training.

Evidence on 'overeducation' or 'underutilisation' of skills is difficult to collect in an internationally comparative context and over time, but organisations across Europe as well as in the US would seem to have considerable margin for enhancing the contributions their staff could make (Tables 4.8 and 4.9). The picture is related to underutilisation of the skills of older workers and of women seeking to return to the labour market. This is apparent, in the case of the former, from the tendency to externalise the employment adjustment required in the face of changing trading conditions through resorting to redundancy and early retirement rather than redeployment. In the case of women, it is apparent from the difficulties they often face in returning to the labour market without passing through a stage of downward mobility in which they are obliged to work at a level significantly below their competence or in another type of occupation not requiring their training.

A. Unemployment (%)*a*

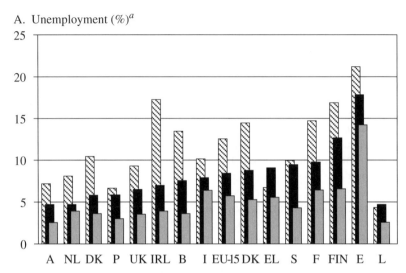

Note: 25–59 year olds (countries sorted by unemployment rates ISCED 3 level).

Source: European Labour Force Survey, 1996.

B. Long-term unemployment by educational attainment, EU-12, 1994 (%)[1]

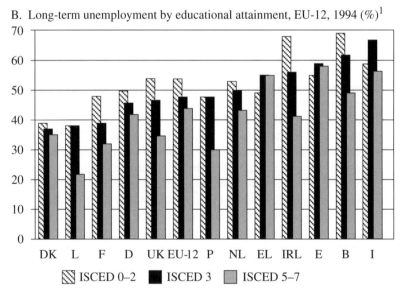

Note: % of total unemployment, 25–59 year olds (countries sorted by ISCED 3 level).

Source: Labour Force Survey, 1994.

Figure 4.1   Unemployment by highest educational/training attainment: EU Member States

Table 4.7   *Expected Years of Unemployment over a Working Lifetime by Level of Educational Attainment for Men aged 25–64 (1995)*

	Below upper secondary	Upper secondary only	Tertiary	Difference in years between tertiary and below upper secondary
A	1.6	0.9	0.6	1.0
B	3.0	1.4	0.9	2.1
CH	2.3	0.9	0.7	1.6
DK	4.0	2.8	2.0	2.0
D	4.5	2.3	1.6	2.9
EL	1.8	1.7	1.9	−0.1
E	5.6	3.9	2.9	2.6
FIN	6.8	5.8	3.1	3.7
F	4.4	2.5	2.1	2.3
IRL	5.0	2.3	1.4	3.6
I	2.2	1.4	1.8	0.4
L	0.7	0.6	0.1	0.6
NL	1.9	1.1	1.1	0.8
NO	2.2	1.4	0.9	1.2
P	1.9	1.6	1.4	0.5
S	4.3	3.3	2.0	2.4
UK	5.4	2.9	1.6	3.8
USA	3.0	1.7	1.1	2.0

Note: The method is extremely crude, being based on current age-education-specific unemployment-population ratios.

Source: OECD 1998, p.110.

4.5   'Human resource regimes' to reduce social exclusion

The above considerations point to an alternative strategy that might benefit the least advantaged more than would targeting resources upon their own training. This is to concentrate on training at the intermediate occupational level in order to remove skill bottlenecks (Lindley, 1991). This would reduce the extent of educational crowding out because those with the ability to do so would train or re-train at the intermediate level rather than filter down, displacing people below them. It would also help to generate more jobs elsewhere in the occupational structure by rendering possible more growth than would have been possible under a skill shortage regime.

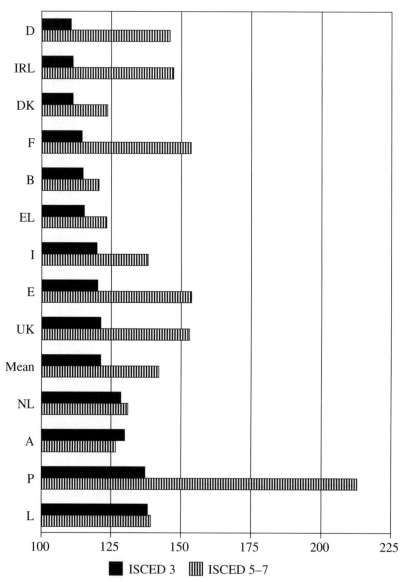

Note: Average net income for people aged 16 years and over in employment (over 15 hours per week). Income was converted to a comparable basis by dividing by purchasing power parities. Amounts are annualised and relate to 1994 with ISCED 2 level = 100.

Source: West and Hind (2000).

Figure 4.2 Relative income according to educational attainment, EU Member States, 1996

Table 4.8   National studies on the extent of over-education

	Year of observation	Study	% of respondents believing they are over-educated	Notes on methodology
D	1992	Pfeiffer/Blechinger 1995	24	self-assessment; apprenticeship trained workers
D	1993	Büchel/Weisshuhn 1996	20	self-assessment
E	1990	Beneito *et al.* 1997	28/15	self-assessment/objective approach
F	1995	Forgeot/Gautié 1997	21	objective approach; 18–29 year olds
NL	1995	Borghans/Smits 1997	22	self-assessment
NL	1995	Hartog 1997	26	comparison with job titles
P	1991/92	cit. by Hartog 1997	33	based on job analyses
UK	1986	Sloane *et al.* 1997	31	self-assessment
UK	1991	Sloane *et al.* 1997	13 (m), 10 (f)	objective approach

Source: Tessaring (1999). The notes on methodology are as provided by Tessaring. Such data may be interpreted in various ways but they indicate (probably except for older workers) that the over-qualification thesis needs to be taken seriously.

At the same time, however, there are labour market and training measures which would help to combat social exclusion through more direct assistance to vulnerable groups. This is not to say that such measures are the most significant ways by which to reduce social exclusion; other broader political and social initiatives are also needed if the conditions of those most affected are to be ameliorated. But several positive steps could, nonetheless, be taken to combat social exclusion via the labour market.

The first is to seek to increase opportunities for higher quality output and jobs in sectors hitherto associated with rather poor conditions of employment: textiles, tourism, and hotels and catering would be examples of this. Support would be necessary through developing training to sustain the emerging occupational structures required to facilitate this strategy.

The second measure is to seek to reduce the risks faced by those who

Table 4.9   Education and training in relation to current job, EU 1995

Member state	% of individuals reporting that formal education and training has contributed to their present work a lot or a fair amount	% of individuals reporting that they have skills for a more demanding job
D	87	65
UK	–	64
A	87	62
EL	98	60
DK	95	60
B	89	60
P	98	54
IRL	94	53
E	72	53
F	94	51
I	85	51
L	92	44
NL	79	40
Mean	89	55

Source: Tessaring (1999), citing West and Hind (2000).

must operate within very tight family budgets with no room to cope with uncertainty, for example, concerning eligibility for social benefit payments or training grants if they were to become involved in training or work experience schemes.

Third, there is a need to provide sustained support through the employment services helping individuals to develop back-to-work strategies and providing the resources needed for effective job search.

Fourth, there is a need to get the policy balance right in a number of related areas, bearing in mind the particular perspective of the most vulnerable groups. These include the balance between:

- providing education as opposed to training;
- initial as opposed to continuing training;
- off-the-job as opposed to on-the-job training;
- training via the internal labour markets of companies rather than the external training market with courses being more accessible to those without jobs;
- and training for the employed as much as for the unemployed (see the example of intermediate skills discussed above).

However, the above strategy which is basically now being pursued by many member states can often distract our attention from the key factor of *cumulative* disadvantage in fostering social exclusion and the fact that this is usually expressed in *local and collective* terms not in *global and individualistic* terms. Moreover, the chronic and persistent nature of social exclusion when it does occur raises questions of balance between the use of *regulatory as opposed to fiscal measures* in different market and non-market contexts in order to achieve significant improvements.

Thus, regulating for rights and responsibilities in the labour market (Lindley, 1993) may serve the interests of both efficiency and equity but there is a potentially more effective way of delivering key social policy objectives for the socially excluded. This is, first, to step partly outside the market and switch from regulatory to fiscal policy instruments concerned with such matters as basic incomes and the quality of general educational opportunities for families and individuals. The second is to work on the boundaries between market and non-market, and to seek to meet both unmet social need as well assisting the transition of people back into mainstream employment. This is a matter for local socioeconomic development and the use of forms of social enterprise and local partnerships. In the knowledge-society context, moreover, attempts to ensure that the socially excluded do 'get connected' need to be seen not, initially, in terms of ensuring that such households have PCs and internet access but in much less atomistic terms. 'Social connection' will be needed to provide the resources needed to sustain an effective internet connection, not the other way round.

In essence, the relationship between skills and social exclusion is much less direct than we would like to be the case. The prospect of tackling exclusion through investing in human capital within the framework of existing labour market and training market structures might appear at first sight to be the most straightforward approach. It is tempting to update this by simply adding another investment component dealing with communications and the internet. However, a significant part of the problem derives from the structures themselves. Reform must, therefore, proceed on a broader front if social exclusion is to be tackled seriously in a period of still high unemployment against the background of the emerging knowledge economy and the unfolding of more ambitious designs for a deeper and wider EU.

5. POLICY DEVELOPMENT AND THE KNOWLEDGE BASE

The implications of the knowledge-based economy are that it will have pervasive effects not only on those sectors that have already been identified as being in the forefront of the development of the information society. It will also affect most if not all other sectors and will have a profound impact upon society at large. Indeed, whether threats are turned into opportunities and whether opportunities are then grasped will depend enormously on achieving a *widespread* appreciation of what is going on. There is a need for a mutually reinforcing relationship between pressures from the demand side as well as supply side in the markets for knowledge, learning and work.

Perhaps, most fundamentally of all, producing the more positive outcomes will depend on how far people first *look to their own practice* rather than advocate changes to that of others. This applies to the policy development system as much as to the parts of the economy and society it seeks to influence. This section deals with the implications of the knowledge-based society for the policy community and for the framework within which policy and practice might evolve together in the education, training and employment policy fields.

Here we are concerned, first, with how the policy community organises itself so as to understand the ways in which the environment for work and learning is changing and what range of future scenarios needs to be explored; and, second, with the sharing and scrutiny of performance in relation to policy experience and the development of progressively more rigorous methods of evaluation.

In the light of the subject matter of the previous sections, three issues are particularly important. The first is the need to pay special attention to the activity of the services sector where the potential for employment growth is concentrated; the second is the need to adjust our thinking so as to decide upon the relative importance of different perspectives on the knowledge-based economy and how this should affect the policy process; and the third is to focus much more on the situations of the socially excluded and the wider group of less well-placed people in society when assessing performance.

The main points made below are necessarily expressed without much discussion or qualification in a paper of this kind but it is worth stressing at the outset that the neglect of some of these apparently technocratic matters does represent a problem generally for the EU if it is to put a knowledge-based policy development system into effect.

5.1   Monitoring and assessment of where the economy and labour market are going

The economic information system is under major strain

- The central information system is ageing as well as the population. Its degeneration has been greatest in the areas where there is most uncertainty about the nature of the socioeconomic change taking place and about its implications for work, learning and social exclusion.

- It is ironic that, at the very time when more accountability and auditing is being introduced by governments to monitor the performance of public services, regulated industries and competition policy at the micro levels of the economy, their ability to measure the volume and quality of output and labour input (and hence productivity) at the macro and meta levels is proving to be so inadequate.
- New approaches to measuring economic activity in services and to the definition of the 'production boundary' which determines what is officially counted as 'economic activity' must be explored in order to retain an adequate picture of what is happening to the economy and society. The debate has barely begun as to how this should be done and how we should use the resulting knowledge base more effectively to develop policy suited to the new socioeconomic environment. Surveys of time use will need to be carried out more intensively and, along with surveys of opinions of services used and their quality, their results will need to be incorporated into the basic data collection system underpinning the national accounts statistics.

The need for more sophisticated monitoring instruments concerning knowledge, skills and social exclusion

- Although there are periodic research projects that study skill acquisition in relation to competitiveness and poverty, the indicators used in performance assessment are much too aggregate and not well-designed to pick up crucial features of either investment in human capital or social exclusion. (Contrast this with the enormous amount of effort devoted to monitoring every facet of the assets and activity of the agricultural sector.)
- At the same time, the introduction of performance indicators in the public sector has embodied a very narrow view of the role of information in society, the responsibility for providing it, and the underlying economics governing its provision. The basic reason for the state introducing such indicators is not because these services are *paid for by the public* but because *they are important to the well-being of the individual, family and community and present major problems for the individual client in assessing the quality of the service.*
- In this context, there is a regrettably supine treatment of the business sector in the employment and training fields. If performance indicators are here to stay, they should be applied to the private as well as

public sectors in the employment and training arenas. In particular, job seekers lack information on the quality of jobs and training opportunities, including the records that different employers and educational and training organisations have in the various areas of human resource development and their different philosophies of 'people management'. The aims of achieving efficient markets for labour and learning cannot be realised without balancing more effectively the information that the organisation expects to have on the individual with the information that the individual can acquire about the employing or educating/training organisation.

- The need for such information on the part of current and potential employees is compatible, moreover, with the needs of shareholders and other investors in seeking to value businesses that are highly knowledge based and hence heavily dependent on the quality of organisational design and implementation.
- Thus it is time for monitoring to take over from *ad hoc* research in a relatively new area. This concerns the progress in creating high performance organisations and displacing poorly performing organisations. This is a potentially controversial area but knowledge of how well countries are *deploying* the labour force is just as important as that of how well they are *educating and training* it.

Regular assessments of the markets for work and learning should be part of the routine monitoring system

- An elementary application of the economics of information points to the need for Europe to adopt the practice of the United States in committing itself to the public provision of routine industrial-occupational projections of one kind or another. But they need to be embedded within broader labour market assessments and complemented by similar efforts to represent the ways in which the education, training and professional development systems are behaving. This is one state aid to industry which would clearly supply a public good that will not be provided by the market at a level of quality that suffices.
- There is too much of a tendency for *ad hoc* working parties, committees of enquiry and opinion polls of the policy and practitioner communities to replace what should be the organised provision of well-digested, routinely produced assessments of the markets for labour and learning.
- The uncoordinated efforts of the various parts of the European Commission have not yet been able to generate best practice in this

area. But this is partly because most member states have, themselves, failed to implement at the national level the necessary monitoring instruments and the analytical capacity required to use them. The proliferation of *ad hoc* studies produced by teams of research institutes that form and re-form in order to bid for one competitive tender after another is no way to exploit effectively the capacity of the European research system and will not yield an adequate substitute for producing well-designed instruments and the organisational structures capable of deploying them. A visit to the US Bureau of Labour Statistics web site will show just how far Europe is behind the US in the provision of information for participants in the various markets for work and learning (see also Goldstein, 1999).

- However, more generally, the European market for information and research about work and learning is deeply distorted by the different conventions which arise in the higher education and research systems in the member states. The frameworks for both EU contract funding and science funding have so far failed to encourage the right balance of cooperation and competition among researchers and consultants and to meet the need for continuity of effort in areas where the creation and dissemination of knowledge requires special attention. This raises issues both of the integrity of the single market in information, consultancy and research services and of the role of EU science policy in the social science and management research fields. These issues need to be addressed.

- The balance of effort devoted to the various European 'observatories' and related information-processing activities is, from the above point of view, too much concerned with monitoring detailed policy reforms than with the labour and learning systems they are supposed to affect. The disparity between the level of routine comparative policy documentation, codified in very precise terms and the crudity of the indicators used in the EU processes is especially striking.

5.2 Sharing and understanding policy experiences

There are different levels at which the experience of member states can be shared and understood: the overall choice of policy strategy, the designs of measures under specific policy areas, the delivery methods used to implement the measures, the systems for monitoring, evaluation, etc. Here, mention will be made of just two aspects: benchmarking and evaluation.

Benchmarking is best used to denote the analysis of performance in relation to practice, followed by reflection, leading to the modification of practice to enhance performance, and so on. It is, essentially, a learning process and is meaningful especially in the context of work groups or organisations which seek to identify what might be considered 'best practice(s)', to explore its nature and then pursue better performance. Useful applications of the technique can be found in various parts of the labour market policy system, notably in the sharing of experience and performance between employment placement offices in the same country (Sjöstrand, 1998) and between employment service of different countries (Arnkil, 1998).

However, the term is so much in vogue that it is also being used for comparisons of policies and monitoring of progress in the absence of clear notions of *practice*. The OECD 'benchmarking policy progress' activity (Elmeskov, 1998) monitors progress towards meeting a set of OECD recommendations given for each country on the basis of a particular view of its structural needs. This cuts across the notion of learning about and evolving best practice. It seems to encompass too much to encourage learning at the micro level because transmission mechanisms for policy implementation are not reviewed, i.e. the policy measure and delivery system are not distinguished. At the same time it encompasses too little to encourage learning at the macro level. This is because there is no conception that, for example, different employment systems might be equally compatible with high performance and that these should be articulated. The process seems to be too wrapped up with the vindication of the Jobs Strategy to generate 'best practice' insights attached to the different approaches. 'Best practice' is the strategy and is subsumed within the country-specific recommendations based upon it.

As regards *evaluation,* the quality of this varies greatly across policy measures and countries. The need for generic evaluation in which findings are suitably codified so that their insights have an application beyond the immediate context of the evaluation has been proposed by Lindley (1996) in the case of the European Social Fund. The principles of this are, however, relevant to several areas of policy: a rigorous codification of *ex post* evaluation findings, which can then be used in *ex ante* evaluation of new proposals, combined with the requirement for progressive improvements in the national/local systems for *ex post* evaluation, while concentrating the EU-funded evaluation effort so as to fill major gaps in the evaluation research base so that collective learning can proceed most efficiently.

Member states are still some way from participating in this kind of process and there has been a loss of opportunity to cut down on unnecessary experimentation, especially with policy measures that have clearly not worked in most environments. This raises fundamental issues which concern the governance of the economy and society and whether or not the citizen has the right to assume that policy will be based on evidence not just on democratic principles. Certainly, there are commitments to share experience and seek out knowledge of which policies succeed in which contexts. But they need to lead to sharper evaluation frameworks and less tolerance within the policy community of poorly designed schemes which essentially use public money to experiment with peoples' experience of work and learning.

6. CONCLUSIONS

So, after all, what is the knowledge-based economy, how does its development alter the European employment debate, and are there modifications that should be made to the EU 'processes' that would help to monitor performance in the light of it? These questions are addressed in this concluding section.

6.1 The meaning of the 'knowledge-based economy'

The use of the term 'knowledge-based economy' or one of the other variations on that theme is bound to invite criticism because knowledge is involved in virtually all activities and the privileging of a group of sectors or a new phase of economic development by referring to them as being 'knowledge-based' seems presumptuous, if not actually ignorant of history. Perhaps, however, use of the term can be clarified, if not entirely legitimised, by allowing it to cover four complementary and in part overlapping phenomena which are growing substantially or should do so:

- the conscious recognition of the notion of 'practice' across a much wider range of occupational contexts than hitherto both at the professional level where it has been less developed (e.g., scientist, computer programmer as opposed to doctor, teacher and engineer) and at other levels (e.g., secretaries, care assistants, nurse auxiliaries, security guards, plumbers, sales assistants);
- the much greater codification of knowledge so that it is accessible and can be better used to bring 'what is practised' into closer association with 'what is known';
- the increasingly diverse production and exploitation of knowledge in more intensive and pervasive ways in many parts of the economy and society *and* the application of knowledge and advanced information processing to the knowledge production and dissemination processes themselves;
- the evolution of a knowledge-based policy development system with very high standards in the collection and use of evidence in the policy process.

This is not to say that a majority of economic life will be dominated by 'reflective practitioners' generating waves of knowledge-based services in a continual state of flux and feedback. But the creation of a knowledge-based economy will be the key innovation facing economic and social life at least in the industrialised nations. Whether this definition is likely to irritate the critics of the metaphor more or less than a vaguer interpretation is a moot point.

6.2.   Knowledge-based economies and the European employment debate

The potential impact of the knowledge-based society will be pervasive. Rights and responsibilities will need to be re-drawn. New understandings will be required, including those forged between social partners in industry, between local partners, and among communities in areas of acute social and economic disadvantage. Partial visions of the knowledge society, focusing only on competitiveness, will be inadequate. They risk losing major opportunities, where virtuous circles of rising productivity, profitability, employment and job quality could be entered and where the pursuit of equity and efficiency could actually reinforce each other. An holistic approach to policy is required.

Consideration of the potential impact of the knowledge-based economy should give more prominence in the employment debate to the quality of organisational design and implementation. These need to be addressed much more explicitly in national and European-level debates which still tend to be more concerned with education, training and skill shortages. However, this does not mean an exclusive concern for the performance of internal labour markets and the support of corporate human resource policies. The knowledge-based society will increase the complexity of interaction between education, training and mobility within organisations, networked employment structures and the external labour market.

The level of investment in education and skills will grow as life-long learning becomes more extensive, especially given the importance of raising the participation of older workers in the labour force (see Esping-Andersen's chapter) and strategies for realising the potential of older workers (Lindley, 1999). One of the main elements of the latter is to retain such workers within the organisation rather than face re-integration problems after redundancy or early retirement. Much of the education and training made available to these workers will thus be in the context of an employment relationship.

In these circumstances, of rising human capital investment both in absolute terms and in relation to other investments, the allowable 'subsidy regime' will need to be carefully worked out; what are, in effect, state aids in the form of direct provision and/or financing to individuals and organisations for education, training and career development will have to come more fully within the scrutiny of the European Commission.

This focus should now take over from the substantial initiatives concerned with European-level regulation of the labour market during the last decade or so. The regulatory 'floor' has been established and, whilst enforcement and compliance require attention, the EU seems to have reached the reasonable limit of its scope for general regulation of the labour market. The evidence does not suggest that *de*-regulation relative to the standards set at the EU level would offer any great pay-off (indeed, it would probably have destructive effects on the attempts to create many more high-performance organisations); this is the case even if some individual countries do have reason to reconsider existing measures that go well beyond the provisions implemented at European level.

6.3.   Proposed modifications to the EU 'processes'

The EU has already taken major steps towards creating a framework for assessing the progress of member states towards the adoption of appropriate macroeconomic, structural and labour market policies. This consists of the initial establishment of the Broad Economic Guidelines and their review at the Cologne Council meeting, together with the introduction of the Cardiff and Luxembourg processes, respectively.

The development of the knowledge-based society does not provide grounds for creating new processes but does suggest modifications to the *content* and methods of scrutiny adopted. There has already been considerable discussion amongst the member states and the Commission about the relationships between the processes and their potential future roles; this will not be pursued here. However, whatever happens to the division of responsibilities arising in the future, the analysis of this paper suggests that some modifications to the content of the processes taken collectively would be desirable. These would include the following five innovations:

- Taking a Rawlesian view of acceptable inequality which deems inequality appropriate only if disadvantaged people benefit from its existence, a key aggregate indicator should be the income accruing to, say, the poorest one-third of individuals or households. In terms of overall performance this would represent the outcome of the various attempts to pursue efficiency and equity. The use of aggregate GDP statistics does not offer any insight into this crucial aspect of performance.
- However, *social exclusion* results from being *trapped* in a state of

cumulative disadvantage. In order to identify how much this is the case, the above indicator needs to be supplemented by one which measures the extent to which the people identified in a given year as being among the poorest are regularly found in that situation or to a large degree move on to better circumstances. It is quite practicable for such indicators to be produced now, even though there will be a need to treat them with care and subject them to refinement.

- The guidelines should be extended to deal with the interfaces between education, training and the labour market. Indicators of changes in the volume of participation in and value of investment in education and training need to be incorporated into the routine monitoring and review processes.

- In dealing with the work and learning aspects of the performance appraisal, there should be a conscious attempt to look at the coherence of the 'model' being pursued explicitly or implicitly by the member state concerned, i.e. the relationship between overall labour market regulation, coordination and social benefits; between market and non-market strategies for investing in education and training; and between the innovation system, the extent of social partnership and policy *vis-à-vis* the encouragement to develop high-performance organisations.

- Additional guidelines should be introduced so that the Commission includes, within its scrutiny of the progress on policy, an assessment of the current quality of the audit, monitoring, evaluation and policy research systems of member states and their plans for improving them. There should also be a report from the Commission or an independent body on the quality of its own systems in this respect.

POSTSCRIPT

This postscript is a highly selective reflection on the outcome of the Portuguese Presidency of 2000. The foregoing chapter has pointed to the likely pervasive effects of the evolution of the knowledge-based economy and society. Unless politicians and the policy community at large absorb the full implications of the 'knowledge-based' metaphor for their *own* work, they will fail both *to develop an appropriate vision for European society and to give substance to it in reality.*

The Portuguese Presidency may, itself, thus be regarded from the point of view of the evolving knowledge-based society. At the same time, this is not the moment to assess the success or otherwise of the Presidency. That is a role for others who come later to the task, where time will lend its own perspective and offer room for analysis that is not feasible in the short term. Meanwhile, this postscript deals with four aspects of the debate and the policy development process initiated by the Portuguese prime minister, António Guterres, at the turn of the new millennium:

1. the orientation of the European 'project';
2. the explicitness of the policy goals adopted;
3. the credibility of the strategies that are put forward to achieve those goals at different levels – the Union, member states, regions and localities;
4. the reliability and effectiveness of the mechanisms for monitoring progress towards those goals and for evaluating the outcomes attained relative to the resources used to achieve them.

The orientation of the European 'project'

After the completion of most areas of the Single European Market followed by the creation of the single currency involving most member states, it is not surprising that there was felt, in some European quarters at least, to be a need for a renewal of the vision of the European 'project'. Moreover, this needed to go beyond that of enlargement which would require yet again lengthy periods of negotiation remote from the lives of ordinary citizens. The production of the 'Competitiveness' White Paper which had sought to regenerate the economic and social debate in the early 1990s had been overtaken by yet more technological change, more intense internationalisation in the form of so-called 'globalisation' and a much sharper realisation of the implications of population ageing. Coming on top of this accumulation of momentum for change were the succession of scenarios embodying progressively more fundamental impacts of ICT on the economy and society at large.

Grasping the millennial moment, the Portuguese government launched its Presidency of the EU with a document that called for 'a new vision and a long-term strategy'.

> A new strategic goal needs to be defined for the next ten years: to make the European Union the world's most dynamic and competitive area, based on innovation and knowledge, able to boost economic growth levels with more and better jobs and greater social cohesion.

Moreover, it asserted that it was essential that 'we regain the condition of full-employment, geared to the needs of the emerging society'. The Lisbon and Feira European Councils endorsed and developed these views inviting the European Commission to take into account various proposals for both policy and processes, especially in the 'labour and learning' fields and including changes to the Employment Guidelines to govern the 'Luxembourg process' for 2001 and beyond.

Underlying the approach is a recognition that the USA is substantially ahead in developing its knowledge-based economy, but this does not mean that Europe has to copy the USA in order to catch up. It seeks to explore different ways of relating the economy to society, exemplified not only by the European welfare systems but also by the commitment to social partnership as a means of conducting relationships between employees and companies and to their deeper joint involvement in the European Employment Strategy and in the conduct of the Luxembourg process. It also seeks to explore further ways of reconciling personal and professional life, for both women and men. It points to the need to examine more extensively the scope for harnessing broader values which blur the boundary between what is 'economic' and what is 'social' by stressing the importance of corporate social responsibility and of finding a more substantial place in the overall strategy for the promotion of sustainable social economy organisations.

This mixed strategy is undoubtedly essential if Europe is to achieve prosperity with diversity and opportunity but without extensive social exclusion. The knowledge-based society calls all the more so for this kind of orientation.

The explicitness of the policy goals adopted

One of the most striking things about the outcome of the Portuguese Presidency is the adoption of more explicit goals and their incorporation into the Employment Guidelines. These include, for example:

- halving by 2010 the number of those aged 18–24 with only lower-secondary education who are not in further education and training;
- achieving 10 per cent of the adult working population (aged 25–64) participating at any given time in education and training by 2010;
- increasing investment in human capital overall by 50 per cent;
- raising the employment–population rate from 62 per cent in 1999 to 70 per cent in 2010 (with that for women rising from 51 to 60 per cent);
- achieving full employment by 2010, defined as equivalent to a 4 per cent unemployment rate;
- reducing the poverty rate from 18 to 10 per cent by 2010 (see Esping-Andersen for discussion of this objective).

Alongside such quantified goals, there are many others which relate to learning and employment, some of which are quantifiable, but this has not yet been done, and some of which are intrinsically difficult to quantify. They include:

- promoting the growth of service sector employment, including personal services;
- improving the basic information and advisory infrastructure of the employment services;
- giving 'higher priority to life-long learning as a basic component of the European social model' and encouraging more progressive ways of organising work so as to foster such learning (the Lisbon Council stipulating that 'Progress towards these goals should be benchmarked');
- the further promotion of equal opportunities, 'including reducing occupational segregation . . . and in particular by setting a new benchmark for improved childcare provision';
- lending greater urgency to the task of reforming social protection so as to achieve the twin goals of raising the EU's employment rate and thereby reducing the extent of chronic exclusion from the labour market and of tackling the poverty of the most vulnerable groups.

As regards the last of these, it must, however, be said that there is still a tendency to choose rather larger 'target groups' than is desirable from the point of view of setting priorities and to see them as demographic categories rather than as different categories of the clearly socially excluded. For example, the Lisbon Council refers to 'the elderly' as a 'specific target group', whereas, in order to set priorities, policy makers need to distinguish more sharply within that group, identifying those who are nearing the latter stages of a life cycle that may have started with child poverty from which they were never able to recover.

However, whilst there is still much to be done, significant progress has been made as a result of the Portuguese Presidency and the cooperation of the various European institutions and fora in thinking through and stating aims that have substance and can be incorporated into the Luxembourg process.

The credibility of the strategies that are put forward to achieve those goals

The consistency of the strategy lies in its aims to raise the potential productivity of all sections of the population of working age and to promote higher and extended labour force participation in the long run, in order to provide resources to deal with child poverty now and have sustainable pension systems in the future. In this scenario, however, the use of education and training can be grossly oversold as the key ingredient in the solution to early disadvantage that continues into working-age experience as if its effects can then, in some sense, be 'neutralised' by a judicious combination of human resource investment, labour market schemes and social protection measures. There is then a danger in being deflected from pursuing the reduction of child poverty with the ruthlessness it deserves (see Esping-Andersen on this point from the welfare perspective).

In addition to keeping a clear sense of proportion about what investment in human resources can deliver, there are also questions about how to deliver the investment in the first place. The commitments to investment in ICT infrastructure (and much wider access to it), R&D, and human capital are clearly essential. However, their ultimate impact on productivity, output and employment will depend on the effectiveness with which organisations use the resulting resources. There is a tendency for the policy system to give less emphasis to the quality of organisational design. The four pillars of 'employability, entrepreneurship, adaptability and equal opportunities' used in the Luxembourg process have been retained in the new Employment Guidelines for 2001. But they have been buttressed, at the request of the Lisbon and Feira Councils, by two 'horizontal sections' which are deemed to be relevant for all pillars and deal with the *role of the social partners* and with *lifelong learning.*

These new parts of the structure of the Employment Guidelines are particularly welcome because they, at least, recognise potential areas of major weakness. The credibility of the strategy depends crucially on there being a strong link between the major boost to education and training, on the one hand, and *potential* individual productivity at work, on the other hand, *and* on there being sufficient jobs available in effective organisations so that the higher potential individual productivity can be realised in practice. It also depends on the substantial switch from 'passive' to 'active' measures (moving from out-of-work to in-work social benefits), yielding a higher supply of suitably skilled labour. Demand and supply then rise together, the former faster than the latter, so as to achieve a higher employment–population rate and lower unemployment rate (as a percentage of both the labour force and the population of working age).

For this mutually reinforcing demand and supply shift to materialise, however, there has to be more than a strong link between investing in human capital and increasing potential productivity. Someone has to *act on their belief* in the link and invest. Europe still seems to lack the mechanisms for leveraging as much investment from governments, private sector organisations and individuals and their families as is envisaged by the strategy to realise the best knowledge-based society in the world.

The mechanisms for monitoring progress and evaluating the outcomes attained

The production of better indicators with which to accompany the Luxembourg process has been progressive and fruitful in several areas but there remain significant weaknesses. Here, just three will be mentioned. First, there is a tendency to understate the return on some basic education, social welfare and health programmes, simply because the benefits are recouped in the longer term and may be rather intangible whereas the costs are incurred in the short term and are relatively easy to quantify. The macroeconomic equivalent of this is the treatment, in the national accounts, of all such social investment not as investment at all but as current spending. The same is true of all other expenditure on education and training, however vocationally related it might be and regardless of the (usually *ad hoc*) evidence of sometimes high social as well as private rates of return. This will produce an increasingly serious distortion of the underlying investment calculus of the growing knowledge-based economy.

The second weakness in monitoring is in the area of changing work organisation. We have very poor European-wide information on this aspect. This has been highlighted in the context of the Joint Employment Report where part of the problem is the poverty of the monitoring instruments but this would seem to be itself a reflection of the fact that there is still too little explicit attention paid to the inadequate quality of organisations in the European economy.

Third, the chapter has referred to the concept of benchmarking and this notion arises in some of the conclusions of the Presidency. The contribution by Lundvall and Tomlinson is devoted to this subject so only brief reference will be made to it here. The main points are that benchmarking must be targeted on particular processes in business, administration, etc. and related to the behaviour of specific actors. The European institutions seem to be developing a penchant for using the term in situations where one or other or neither of these conditions apply.

At the same time, as argued in the conclusions to this chapter, we are still not codifying sufficiently the different overall models whose characteristics are more or less present in particular member states or in hypothetical variants based upon them. This deals with the relations between the parts (notably the innovation system, labour market regulation, social benefits, economic coordination, market and non-market strategies for investment in human resources, and the place given to social partnership) rather than their individual performances. The Luxembourg process is yielding a valuable framework for policy development and exploiting the diversity of past experience and current experimentation found among member states. The proposals of the Lisbon and Feira European Councils, combined with the Mid-Term Review of the Luxembourg process, ensure that the framework will continue to evolve. But it does need a 'meta-level' at which the broader vision and long-term strategy initiated by the Portuguese government in January 2000 and subsequently taken up by and developed with its European partners can be debated more directly and vigorously. Only that way will Europe really evolve a model or models that bring the full benefits of the knowledge society to all of its citizens.

ACKNOWLEDGEMENTS

This chapter was originally one of four background papers prepared for the Portuguese Presidency of the European Union under the action line 'For a Europe of Innovation and Knowledge with More Employment and Social Cohesion' and distributed with the Presidency document in January 2000. The present version contains some minor amendments and a Postscript written after the end of the Presidency. I wish to thank the Portuguese Government for inviting me to participate in the activities concerned with their Presidency. I have especially appreciated the interest which the Prime Minister, António Guterres, and his fellow ministers showed in the work and the opportunities provided to exchange views about the issues covered. I have a special debt to Professor Maria João Rodrigues of the Prime Minister's cabinet for her advice and comments during her coordination of the work of the Group. In addition, I am grateful to Robert Boyer, Gøsta Esping-Andersen and Luc Soete, who prepared the other three initial background papers, for discussions with them in the course of preparing my own paper; to Mario Telò and Bengt Lundvall for subsequent discussion arising out of the book project; and to Alan Brown, Sheila Galloway, Manfred Tessaring and Rob Wilson for comments on particular issues or for allowing me to use statistical material produced by them. However, responsibility for the views expressed lies entirely with the author.

REFERENCES*

Arnkil, R. (1998), 'Benchmarking employment services: international comparisons', in European Employment Observatory, pp. 112–20.
Boisot, M.H. (1999), *Knowledge Assets: Securing Competitive Advantage in the Information Economy*, Oxford: Oxford University Press.

Castells, M. (1996), *The Rise of the Network Society*, Oxford: Blackwells.

Commission of the European Communities (1998), *Employment in Europe*, Luxembourg: Office for Official Publications of the European Communities.

Commission of the European Communities (1999), 'Strategies for jobs in the information society', Report to the European Council, Information Society/99/22, Brussels: CEC.

Commission on the Future of Worker-Management Relations (1994), *Fact Finding Report,* US Department of Labour/Department of Commerce, Washington: US Government Printing Office.

Coyle, D. (1997), *The Weightless World*. Oxford: Capstone.

de Grip, A and H. Heijke (1998), 'Beyond manpower planning: ROA's labour market model and its forecasts to 2002', Maastricht: Research Centre for Education and the Labour Market, Maastricht University.

Elmeskov, J. (1998), 'The OECD jobs strategy: Benchmarking policy progress', in European Employment Observatory, pp. 81–90.

European Employment Observatory (1998), 'Benchmarking labour market performances and policies', Joint Employment Observatory Conference Proceedings, Rotterdam, 2–3 July, Berlin: I.A.S.

Gibbon, M. *et al.* (1994), *The New Production of Knowledge*, London: Sage.

Goldstein, H. (1999), 'The early history of the Occupational Outlook Handbook', *Monthly Labor Review*, May, 3–26.

Gray, G.R., D.W. Myers and P.S. Myers (1999), 'Co-operative provisions in labor agreements: a new paradigm', *Monthly Labor Review*, January, 29–45.

Green, F. (1999), 'The market value of generic skills', Skills Task Force Research Paper 8, Sheffield: DfEE.

Grubb, D. and W. Wells (1993), 'Employment regulation and patterns of work in EC countries', *OECD Economic Studies*, 21 (winter), 7–58.

Kelly, K. (1998), *New Rules for the New Economy*, London: Fourth Estate.

Lindley, R.M. (1991), 'Interactions in the markets for education, training and labour: a European perspective on intermediate skills', in Paul Ryan (ed.), *International Comparisons of Vocational Education and Training for Intermediate Skills*, Basingstoke: Falmer, pp. 185–205.

Lindley, R.M. (1993), 'Rights and responsibilities: Contrasting expectations of European employment policies', in J. Freyssinet (ed.), *Convergence des Modèles Sociaux Européens*, Paris: Ministère du Travail, de l'Emploi et de la Formation Professionnelle, pp. 248–59.

Lindley, R.M. (1996), 'The European Social Fund: A strategy for generic evaluation', in G. Schmid, J. O'Reilly and K. Schömann (eds), *International Handbook of Labour Market Policy and Evaluation*, Cheltenham, UK and Brookfield, US: Edward Elgar, pp. 843–67.

Lindley, R.M. (1997a), 'Socio-economic environments and rule-making in the EU', in D. Mayes (ed.), *The Evolution of the Single European Market*. Cheltenham, UK and Brookfield, US: Edward Elgar, pp. 114–66.

Lindley, R.M. (1997b), 'Labour market flexibility and the European Union', in A. Amadeo and S. Horton (eds), *Labour Productivity and Flexibility*, London: Macmillan, pp. 150–71.

Lindley, R.M. (1999), 'Population ageing and the labour market in Europe.' *Rivista Italiana di Economia, Demografia e Statistica,* 53(1), 67–92.

Marshall, R. (1994), 'The Commission on the Future of Worker-Management Relations', First Round Table on the Social Dialogue (Perspectives on the

Employment Relationship: The US Commission on the Future of Worker-Management Relations), Brussels, April.

Marshall, R. (1998), 'États Unis: crise et avenir incertain des relations industrielles: autour des traveaux de la commission Dunlop', in H. Nadel and R.M. Lindley (eds), *Les Relations Sociales en Europe,* Paris: L'Harmattan, pp. 47–71.

Nadel, H. and R.M. Lindley (eds) (1998), *Les Relations Sociales en Europe*, Paris: L'Harmattan.

Nickell, S and and R. Layard (1998), 'Labour market institutions and economic performance', Discussion Paper no. 407, London: Centre for Economic Performance, London: School of Economics.

OECD (1994), *The OECD Jobs Study*, Paris: OECD.

OECD (1998), *Human Capital Investment. An international comparison*, Paris: OECD.

OECD (1999), 'Labour market performance and the OECD jobs strategy', in *OECD Economic Outlook*, Paris: OECD, pp. 47–132.

Reich, R.B. (1991), *The Work of Nations*, New York: Vintage Books.

Scarborough, H., J. Swan and J. Preston (1999), *Knowledge Management: A Literature Review*, Issues in People Management, London: Institute of Personnel and Development.

Schmid, G., H. Schütz and S. Speckesser (1998), 'Broadening the scope of benchmarking: radar charts and employment', European Employment Observatory, pp. 54–70.

Schön, D.A. (1983), *The Reflective Practitioner*, London: Maurice Temple Smith.

Sexton, J.J., D. Frost and G. Hughes (1998), 'Aspects of occupational change in the Irish economy – recent trends and future prospects', FÁS/ESRI Manpower Forecasting Studies Series, Report No.7, Dublin: FÁS/ESRI.

Shapiro C. and H.R. Varian (1999), *Information Rules: A Strategic Guide to the Network Economy*, Boston, MA: Harvard Business School Press.

Silvestri, G.T. (1997), 'Occupational employment projections to 2006', *Monthly Labor Review,* November, 58–83.

Sjöstrand, K-M. (1998), 'Benchmarking labour market performances and policies: Sweden', European Employment Observatory, pp. 121–31.

Stewart, T.A. (1997), *Intellectual Capital*, London: Nicholas Brealey.

Swedish National Labour Market Board (1999), 'Jobs for the future – choosing an occupation into the 21st century', *ISEKen*, 13/98.

Tessaring, M. (1999), 'Human resource potential and the role of education and training', CEDEFOP (Internal Paper), Thessaloniki: CEDEFOP.

US Department of Labor (1991), *Labor-Management Commitments: Compact for Change*, Washington: US Department of Labor.

van Ark, B. and R.H. McGuckin (1999), 'International comparisons of labor productivity and per capita income', *Monthly Labor Review*, July, 33–41.

Veneri, C.M. (1999), 'Can occupational labor shortages be identified using available data?', *Monthly Labor Review*, March, 15–21.

West, A. and A. Hind (2000), *Indicators for vocational education and training: exploitation of the European Community Household Panel ECHP*, Thessaloniki: CEDEFOP. [www.trainingvillage.gr/download/statistique/echp/index.html].

Wilson, R. (ed.) (1998), 'Review of the economy and employment 1998/99', Coventry: Institute for Employment Research, University of Warwick.

N.B. Basic documents relating to the EU processes and their development are not included above and are not cited formally in the text. The results of labour market studies which are drawn upon in summarising key features of occupational change in section 3 are too numerous to include full citations here.

5. Institutional reforms for growth, employment and social cohesion: elements of a European and national agenda

Robert Boyer

> The difficulty lies, not in the new ideas, but in escaping from the old ones, which ramify (. . .) into every corner of our minds.
>
> John Maynard Keynes, December 13, 1935.

> The outstanding faults of the economic society in which we live, are its failure to provide for full-employment and its arbitrary and inequitable distribution of wealth and incomes.
>
> John Maynard Keynes, *General Theory*, Chapter 24, 1936.

1. INTRODUCTION: IN RESPONSE TO THE PRESENT DECADE'S CHALLENGES, IT IS TIME TO REFORM

The recovery of the EU economy has brought a lot of optimism to European business and consumers. Hasn't GDP growth gone up from 1.5 per cent during the period 1991–6 to 2.7 per cent during the following years (1997–2000). Is the European sclerosis over, since the national and European institutions are now in line with the requirements of global finance? This chapter proposes a balanced view: many reforms have already been done, but they are quite unequal across the fifteen member states and in any case new challenges are to be met during the coming years. It argues that since the 1990s the Europeans are living a brand new period that deserves careful and new analyses. Thus, most of the items on the political agenda should be re-assessed and probably redesigned according to three major concerns.

The turning point of the 1990s has pointed out some recurring weaknesses of Europe. It is now clear that the 1990s exhibited intense and multifaceted transformations with quite unequal consequences for the three members of the triad:

- The previous Fordist productive paradigm has progressively led to *alternative principles*, based on a knowledge economy (OECD, 1999). This is a major challenge for Europe as a whole and many member states (see Chapters 2 and 4).
- Nevertheless the dynamism of *financial innovations* has overcome the speed of technological and industrial advances (Boyer, 1999b).
- The so-called *globalisation of finance* and diffusion of export-led growth strategies to Newly Industrialised Countries (NICs) has propagated local financial crises to the rest of the world.
- A large *de-synchronisation of the business cycles* has nevertheless been observed since the early 1990s, with some impact upon the misalignment of exchange rates euro/dollar and yen/dollar.
- More basically, the miracles of 1960s (e.g., France, Germany, Scandinavian countries and even Japan, East Asian NICs) have turned into seeming failures and major crises. Therefore, many experts think that any government struggles to implement *market-friendly institutions*.
- The perception of a *crisis of the welfare state*, specially of its financing, has been triggered by the post-1973 growth slowdown and more recently by the prospect of an ageing population (see Chapter 3).

Most European countries have been adversely affected by these structural transformations. It took a long time for economic and political decision makers to take the measure of the far-reaching consequences of this turning point. Many firms are lagging in terms of the diffusion of information and communication technologies (ICT) and still more in the production of the related hardware and software components. Macroeconomic performance has been poor, especially when compared with the US during the 1990s or past European growth during the 1960s. Last but not least, unemployment has resisted many therapies, at least for medium-sized European countries, until 1995. The legacy of the 1990s in terms of macroeconomic disequilibrium calls for new directions for economic policies. But this is not the only argument.

1.1 Coping with the complete reversal of the post-Second World War institutional architecture

It is now more and more evident that these difficulties have a deeper root than a mere mismanagement of the European policy mix after the German reunification and the costs associated with the preparation for the launching of the euro. Much research suggests that the institutions that were at the core of the Golden Age's unprecedented growth have been destabilised by the very success of this regime and the related surge of new trends within the world economy (Boyer and Saillard, 2000; Aglietta, 1999; Baslé, Mazier and Vidal, 1999). A sharp contrast emerges when the 1990s are compared with the 1960s.

- Clearly, *during the 1960s*, an original capital labour accord was the cornerstone of the diffusion of mass production and consumption, since it was generally organising the quid pro quo between the acceptance by workers of Fordist productive methods and the institutionalisation of real wage increases in line with productivity (Figure 5.1). Basically all other institutional forms were organised in accordance with this core compromise: oligopolistic competition on a national basis was the rule, the state was organising the welfare state along with Keynesian fine-tuning policies, the monetary policy was reflecting an abundant credit supply, at the possible cost of periodic devaluation when inflation was no more in line with world trends. The whole macroeconomic regime was built upon this clear institutional hierarchy.

- *After 1990*, the picture is quite different (Figure 5.2). The leading institutional form relates to the insertion of each economy into the world system, in terms of trade, investment, credit and global finance. Thus, the competitive forces permeate from one country to another, even if they may finally tend to the formation of international oligopolistic groups, as shown by the international merger mania of the late 1990s. Similarly, national monetary policies are governed by the appraisal by the international financial community and thus, the fine tuning of the policy mix is made far more complex. In the early 1990s slow growth has brought recurring and long-lasting public deficits, difficult to finance given the high interest rates implied by the strict monetary policy implemented by central banks and the international imbalance between investment and saving trends. Finally, the capital labour accord is under strong pressure, since it has to cope with the constraints exerted by the stiffening of competition and the emergence of a lean state. The frequency of the call for more labour market flexibility is good evidence of this complete reversal of the institutional hierarchy of the post-Second World War.

This shift might well be one of the underlying and neglected factors explaining why Social Europe is so difficult to implement, but conversely

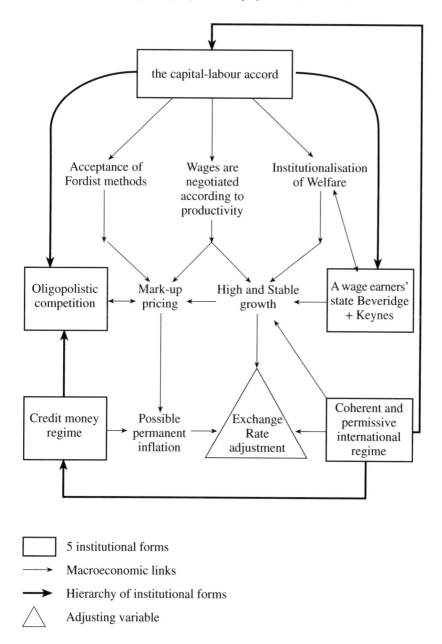

Figure 5.1 The Post-WWII capital–labour accord shaped most other socioeconomic institutions

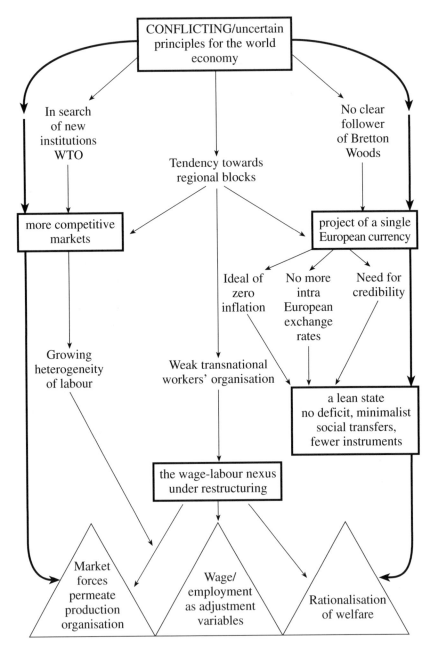

Figure 5.2 The euro implies a new hierarchy and architecture of each national socioeconomic regime

more required than ever (Maurice, 1999). But there are many other difficulties facing the present European institutions after the Amsterdam Treaty and the December 2000 Nice summit.

1.2   European integration: work in progress . . . not at all the end of history

On top of these world-wide issues, European decision makers are facing quite specific issues:

- Does monetary integration mean *the end of the economic and political process* contemplated by the founding fathers of Europe? What compatibility is there with the still independent budgetary policies and national political processes?
- What about the long-run *economic and political viability* of a system that is partially federalist (on, for example, the issue of money, competition, international trade policy) but still deeply embedded in the national political arena, in spite of the growing role of the European Parliament (Boyer, 1999a)? What are the relationships between governance and governments in the European Union (Chapter 8)?
- While foreign analysts tend to blame and fear the so-called 'fortress Europe', is not the Old Continent suffering from *a lack of coordination* among member states, specially in terms of foreign policy and defence?
- Is the European common monetary policy compatible with the *persisting diversity* of national regulation modes, political coalitions, welfare systems and life styles? How can a European identity be constructed in order to overcome the risk of balkanisation (Chapter 7)?
- Last but not least, many experts and political leaders have come to express some doubts about the desirability and long-run viability of the so-called *European social model* in the era of finance and globalisation. Is it not time to give a new definition to this model (Chapter 3)?

The various chapters of this book provide many insights and proposals about these different issues. This contribution tries to relate one issue with another and stresses the need for a coherent institutional architecture.

1.3.   The dilemma of the European Union

The launching of the euro has made the paradox of European integration quite clear. Seen from outside, the Old Continent displays a quite an impressive economy, but, analysed from within, many structural and institutional disequilibria have to be overcome, in order to convert this economic and political strength from the potential to the effective.

1.4.   A potential economic giant . . .

Much more than a wide and overwhelming globalisation the last decades have experienced the emergence of a *triad configuration*, among which Europe is not at all a minor partner (Figure 5.3). More populated than the US, the European Union has a market size nearly equivalent to that of the US and the average productivity level is similar to the Japanese one. The research and development expenditures are more important than those of the Japanese and the number of scientists and engineers roughly equivalent. Nevertheless, given the relative heterogeneity of the member states, the level of productivity is far inferior to that of the US in terms of purchasing power parity, and, simultaneously, the US scientific and technological potential is far superior. But the most severe European problems are elsewhere.

1.5.   . . . in search of a relevant strategy . . .

First of all, even if the Maastricht and Amsterdam Treaties follow the continuity of the process of economic integration launched with the Single Market Act, they imply quite a significant restructuring of *the distribution of responsibilities* between national authorities, the European Commission, the European Parliament and the European Central Bank. Far from being a totally determinist trajectory, the process that has been initiated is open to various new issues, emerging political difficulties and the need for redesigning the policy mix. Many futures can thus be contemplated with quite contrasting outcomes (Boyer, 1999d). Self-satisfaction, dogmatism or inertia in the readjustment of priorities could be quite dangerous indeed.

1.6.   . . . facing severe and new economic and political coordination
problems

A major challenge is still the question of the European employment. If back in the 1970s and 1980s it could be analysed as the consequence of the international crisis and the breaking down of the Fordist growth regime, in the 1990s, unemployment was clearly a European disease, at least among developed countries. In spite of the current recovery, the average European unemployment rate is twice as high as those of America and Japan. This feature is conventionally interpreted as evidence of labour market rigidities

A MAJOR POLE OF THE TRIAD

1999

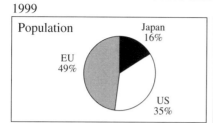

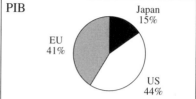

... STILL SOME LAG IN PRODUCTIVITY, SCIENCE AND TECHNOLOGY

Product per inhabitant/purchasing power parity

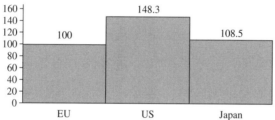

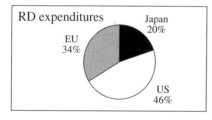

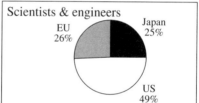

... BUT ON AVERAGE A HIGHER UNEMPLOYMENT RATE

1999

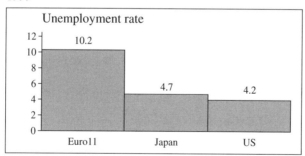

Figure 5.3 European Union: a potential economic giant but a lot of coordination problems

Table 5.1   A comparison of the degree of coordination among the triad

	European Union	US	JAPAN
Monetary policy	Independent ECB	Independent FRB	Towards an independent Central Bank
Budgetary policy	12 budgetary policies constrained by the stability and growth pact	One federal budget, several states budgets	Control by Ministry of Finance
Tax policy	Emerging principles for co-ordination, but not harmonisation	Federal taxes and state taxes differentiation	Homogenous tax system
Quality of the policy mix	Improving via learning process within the informal Eurogroup	A long experience in co-ordinating monetary and budgetary policies	Previously obtained via Ministry of Finance
Wage bargaining	Heterogeneous institutions across nations, sectors, skills: largely uncoordinated process, mixing competition for some and protection for others	Near complete decentralisation, rather competitive forces	Mix of largely firm-based decentralised wage systems and synchronised wage increases
Competition	From national oligopoly to the search for a single market	After the wave of deregulation a trend towards large financial concentration, but large entry of new firms	The structure of the Keiretsu allows both internal coordination and significant external competition
Insertion into the international economy	Single external tariffs and exchange rate but still different national interests	Variety of tools (lobbying, anti-dumping devices) in order to cope with diversity of interest across states	Traditionally, discrete but strong controls over the inflow of goods, investment, manpower

and the absence of strong competition. But it is evidence too that major coordination problems have not been solved. First, there are those between firms, unions and workers, even though by chance strong national counter examples do exist. For instance, the social pacts negotiated in the Netherlands, Ireland and Denmark seem to have been quite instrumental in curbing unemployment. Second are those between monetary authorities and Ministers of Finance. Back in the early 1990s high real interest rates had worsened the public deficit and thereby reinforced the lack of credibility of national government strategies, as assessed by the international financial community. Third, in spite of ambitious TSER programmes, the national strategies for innovation and technology do not display the synergy that might be expected from the European Union. Fourth, the external representation of EU is not fully established and this weakness may have played some role in the surprising depreciation of the euro with respect to the dollar, from January 1999 until November 2000. In one word, Europe is suffering from *a lack of coordination mechanisms* in all the domain of economic policy (Table 5.1). Of course, all the actors are beginning to experience the cost of such failure and they may learn quickly, but it might be useful to help in diagnosing some major issues and their mutual links.

The European dilemma is rather simple: how to foster *growth and employment* while preserving or even extending *social cohesion*? Is there a single best way, or should benchmarking be used as a policy learning tool (Chapter 6)? The basic message of this chapter is that several options are open to the EU. They are not at all equivalent and they call for alternative institutional redesign and economic policy strategy.

2.   STRENGTHS AND WEAKNESSES OF EUROPEAN UNION

They differ significantly according to the various economic paradigms (Table 5.2) that may prevail in the early twenty-first century (Boyer, 1999a).

2.1.   From Fordist mass-production to Toyotism?

If the source of competitiveness were closely related to *differentiation by quality and possibly by innovation* in terms of mass-produced goods, then Japan should still be leading in the production of typical Fordist goods, whereas the US is far from implementing such a model in North America as evidenced by the American trade deficit in typical mass-produced goods, such as cars and conventional consumers' electronics goods. Given the

Table 5.2 *What growth regime for the early twenty-first century?*

Regime\ Economies	Toyotism as a follow-up of Fordism FBCF/GDP (1)	Information and communication technologies-led		Knowledge-based economy (Education + RD + Software)/GDP (3)	'New' service-led economy Employment in the service/ Working population (4)	Export-led growth Share of export/GDP		Foreign competition-led growth Intensity of competition E (6)
		Expenditures in ICT/GDP (2)	Personal computer/ Blue collar worker			1999	Variation 1991–9 (5)	
European Union	19.0	5.8	54	8.0	39.2	10.0	+2.9	27.6
US	17.5	7.6	103	8.4	53.8	11.4	+1.1	29.7
Japan	28.5	6.3	18	6.6	46.0	12.2	+2.0	21.2

Note: Intensity of competition is defined as: $E = X/Y + (1 - X/Y).M/D$ X export, M import, D domestic demand, Y production.

Source: (1), (2), (3), (4): OECD, 'L'économie fondée sur le savoir', 1999.
(5) *Economie Européenne*, n° 66, 1998, pp. 149, 142.
(6) OECD *Perspectives Economiques*, June 1999, p. 237.

diversity of production paradigms in Europe (Boyer and Durand, 1997) and the intermediate rate of investment, Euro15 is not that bad (Table 5.2, first column). But the current difficulties of the Japanese economy as well as a more prospective approach seem to suggest that deeper transformations have occurred during the 1990s and that this mere modernisation of the post-world war regime is not sufficient, even if necessary. Furthermore, the permanent capital deepening (more capital per worker) is no longer associated with major labour productivity increases, and the deterioration of the output/capital ratio in Europe might be one of the origins of a difficult recovery of profitability, hence of the creation and persistence of mass unemployment.

2.2. The rise of information and communication technologies: a shift in the productive basis

The long history of industrial revolutions suggests a second and quite different hypothesis: the emergence and diffusion of such *generic technologies* as information and communication technologies (ICTs) would mean the implementation of a totally new productive paradigm. This is a typical Schumpeterian hypothesis, now more and more widely accepted. In this respect, the competitive position of the EU is quite preoccupying: in spite of high R&D efforts, European firms seem to have been unable to extend their world market share in ICTs (Amable and Boyer, 1993) and this is one of the major weaknesses for Europe (Soete, 1999). The gap between the US and the EU in terms of the absolute amount of R&D has drastically increased during the 1990s (Chapter 2). Many statistical indexes confirm such a lagging position of most European countries (Table 5.2, column 2). On the supply side, the EU is relying on a significant flow of imports from the US and, on the demand side, on average, European firms are investing less than Japanese and American corporations. In terms of diffusion of the use of modern technologies, such as computers by employees, the European situation is intermediate between that of the US and Japan.

Maybe the situation is not as bad as it may seem at first look. First, detailed statistical and econometric studies suggest that the impact of ICTs on growth and employment is positive and significant but the size of the effects are rather modest indeed (Boyer and Didier, 1998). Second, even in the US, the leader in ICTs, job creation in the related industries is important but does not represent a major source of the so-called American miracle in terms of the job-creating machine. Nearly the same hierarchy is observed for job creation for the whole OECD during the period 1980–95 (Table 5.3). Third, some international comparisons show that there is no clear correlation between the mastering of any single emerging technological paradigm

Table 5.3   Net job creations take place in quite diverse sectors, not only in those that are high tech OECD, 1980–95 (Variation in percent)

	%
Collective, social and personal services	65.0
Financial services and services to business	62.5
Rubber and plastic	24.9
Retailing, hotel, restaurants	18.4
Pharmaceuticals	14.6
Computers and related	8.6
Transport and logistics	7.4
Paper and printing	7.3
Electronic components	4.6
Chemistry	3.9
High-tech sectors	3.3
Cars	0.3
Medium – Average technology	−8.2
Low-technology sectors	−10.9

Source: Second European Report on Scientific and Technological Indicators (1997:4).

and unemployment performance (Amable, Barre and Boyer, 1997a). Fourth, no innovation in the contemporary world can be reduced to ICTs, since many other sources of technological and scientific advances are observed and may play a major role in fostering growth.

2.3.   Towards a knowledge-based economy?

Clearly, the mechanisms for diffusing information should not be confused with knowledge creation and use: knowledge may use the channel provided by ICTs for its diffusion, as soon as it can be codified and then digitalised. But ICTs *per se* do not create knowledge outside their own domain. Thus, there is an alternative interpretation of the contemporary transformation in production and marketing organisations: the advances in basic knowledge would be more easily converted into profitable new products by close interconnectedness of *scientific research, market analysis and flexible manufacturing*. This may well define a new paradigm, premise of a knowledge-based economy (KBE). Of course, ICTs do help the efficiency of this paradigm and speed up the diffusion of knowledge, they provide the infrastructure, not the essence KBE. From a theoretical point of view, information and knowledge are to be distinguished. Various statistical indicators have been built to capture the diffusion both of ICT and KBE, and they tend to be somehow different.

Actually, the statistical indexes recently elaborated and published by OECD (1999) tend to show that the situation in Europe is finally close to that in the US, whereas Japanese society suffers from the relative poor performance of its academic system (Table 5.2, column 4). This assessment takes into account the fact that R&D expenditures, public education, software (design and copyright are not presently incorporated into the OECD indicators) are all factors that enhance transferable knowledge, thus the ability to develop new ideas, products, processes and organisations.

This trend toward the abstraction of production and the new sources of economic performance simultaneously generate some forms of tacit knowledge that raise the power of a limited fraction of the employees able to implement the radical innovations that determine the competitive position of a firm: for example, creative people in advertising, financial experts creating new options, software designers, artists at the core of the leisure industry. One might recognise the symbolic analysts pointed out by Robert Reich, highly internationalised and who play a significant role in the widening of income and wealth inequalities. In this respect, the highly regulated institutional context of most continental European countries seems to have been detrimental to the maturation of such a leading group. The European Council resolution on *the 1999 Employment Guideline* tackles this issue and proposes various measures in order to develop entrepreneurship. It has recently been proposed to introduce cross-European mobility, just to fully exploit the potential of a KBE economy (Soete 1999). The frequent reference to the American experience seems to imply that policy makers are facing a trade-off between economic performance and reduced income inequalities, a dilemma that Europe has to overcome. The case of small open economies suggests that this not a fatality and that social pacts may deliver simultaneously dynamic efficiency and social cohesion.

2.4.   Are contemporary economies service led?

But there is still another vision of the future of growth and employment. It starts from the evidence that quite all net job creations take place within the service sector, even if the precise distribution may vary from one country to another. Nowadays, the manufacturing sector experiences a slimming down by the subcontracting of many tertiary activities to highly specialised firms in, for example, accounting, finance, marketing and human resources management. Therefore, many sources of firm's competitiveness are manufactured within the so-called service sector, specially for modern *business services* (Petit, 1998).

Thus, when job creation during the last decades as well as forecasts are analysed (Soete, 1999, Table 2 and Table 3), a quite unconventional picture is obtained, when compared with the clichés about the future of employment. Again the US seems to be leading: according to the Labour Statistical Bureau forecast, employment linked to computer and data processing is supposed to have a 7.6 per cent annual growth rate until 2006. But, simultaneously, all jobs related to *health care* will grow at 4.0 per cent, higher than business and financial services, not to mention environment-related and leisure activity jobs which will also experience fast growth. The same picture for the OECD as a whole is obtained over the period 1980–95 (Table 5.3). Therefore, ICTs and even knowledge-based activities will represent a significant contribution to employment growth. But they will not constitute the totality.

A more radical picture of a service-led growth can be proposed. Many components of the services try to cope with the many differing needs of social life in large urban centres; for instance, care for children within families where both the father and the mother have a job, either full-time or part-time, and day care for elderly people; restaurants, retailers, distribution and transportation services, education services and of course the bulk of health care. Some analysts even think that the transformations related to an *equal gender opportunity* inside the family and within the whole society are powerful enough to engineer a new virtuous circle of changed life styles – consumption transformations, insertion of new talents into the economy, new directions for innovation, and so on (Majnoni d'Intignano, 1999).

According to this interpretation and prognosis, the slow growth of unemployment in Europe can be attributed mainly to the lag in the development of a fully fledged service economy, a statement found in many European reports (for instance, European Commission, 1999: 16). Actually, should the share of services in total employment reach the US level, Europe could enjoy an employment rate 14.6 per cent higher than actually observed. Of course, the situation is quite unequal across member states: fewer services in Germany but highly developed in Denmark, the UK, Sweden and so on. But the question is then: could the Europeans create as many jobs as the Americans do, while preserving relatively high wages and good social protection? Some evidence suggests that the EU has fewer people employed both in the high-wage and low-wage industries (Freeman, 1998).

2.5.   A constant deepening of competition, the basis of a new growth regime?

A different view stresses that European growth is directly linked to the competitiveness *of the Old Continent*, in a world wide open to trade, investment, finance and even some highly skilled professionals. During the long and painful 1970s and 1980s, each European country struggled in order to restore its competitiveness, either by labour cost reductions or by innovation and quality. From a theoretical point of view, the form of competition, among firms and nations, is now governing the structural adjustment in the wage-labour nexus (e.g., employment flexibility, wage moderation, rationalisation of the welfare state), the relative degree of taxation of mobile and immobile factors. Economic policy formation itself is affected. Thus, the recovery of growth has been frequently assumed to be dependent on the emergence of a totally new growth regime: cost and price moderation, increase in exports, spill-over from investment to consumption, and finally recovery in demand. A cumulative growth pattern would result from the interaction of *competitiveness and domestic demand.*

This model is quite relevant for small open economies and has been imposing itself on medium-sized economies during the 1980s. But precisely the constitution of the single market and the creation of the euro do change the respective importance of competitiveness and internal market dynamism. Of course, each firm in order to stay in business has to be profitable, but the vanishing of exchange rate variability and the changing pattern of expectations associated with the European Monetary Union (EMU) give a better chance to a domestic-led growth regime. After all, globally Euro15 is nowadays less open than the US or Japan (Table 5.2, two last columns) and this gives more degrees of freedom to the design of a better policy mix, combining monetary stability, growth dynamism and job creation. All the players will progressively learn the new properties of a common monetary policy and the essential benefits associated with it (Boyer, 1999d; Jacquet and Pisani-Ferry, 2000). Thus external competition could be mitigated by comparison with the 1990s, even though a significant degree of internally enforced competition is necessary to promote innovation and a permanent search for efficiency.

2.6.   Is a finance-led regime possible for the EU?

There is still another interpretation of the macroeconomic divergence between the US and the EU. The US can fully incorporate the consequences of financial innovation and globalisation by transforming the governance mode of corporations, the management of the financial system, by privatising large segments of welfare and by developing pension funds (Boyer, 1999c). This would lead to a totally *novel 'régulation mode'*, currently labelled 'the new economy', which would include: labour market flexibility, price stability, a surge in high-tech sectors, a booming stock market and the availability of credit in order to sustain the rapid growth in consumption and the permanent optimism of firms and financial markets. The poor performance of the EU would be closely related to a large lag in catching up with this finance-led regime. For some authors, this would be the real follower of the Fordist mass production growth period (Aglietta, 1998).

The boom in mergers and acquisitions in Europe after January 1999 seems to confirm such a hypothesis: the mega-mergers within many manufacturing sectors, the constitution of large financial groups, the project of implementing pension funds in order to complement previous systems perceived to be under severe strain due to the ageing of European population, all these factors seem to be evidence for the next convergence of European and even Japanese economies toward such a finance-led regime, already at work in the US.

Nevertheless, a closer look at the organisation of the financial systems of the triad and simple statistical indexes suggest that, with the exception of UK, few European countries are able to enter into such a finance-led regime. Furthermore, some theoretical analyses hint that the exhilarating phase of the diffusion of new financial instruments, which is favourable to growth, may end up in a zone of *structural instability*, by the very success and generalisation of behaviour of businesses and households quite exclusively governed by the optimisation of financial rates of returns (Boyer 1999c). Therefore, the EU should be careful in reaping the benefits of financial liberalisation, without entering the dangerous zone of financial and economic instability. Incidentally, Europeans should be quite active in promoting a viable *new architecture for the international financial system*, that remains quite shaky, in spite of a renewed optimism at the end of 1999 (Davanne, 1998; Boyer, 1999b). The more so after the collapse of the internet bubble after March 2000 and the diffusion of this recessionary shock to the world economy.

2.7.   Each growth regime calls for specific institutional architecture.

This brief survey tends to conclude that each interpretation captures some significant aspects of contemporary structural transformations but that none should pretend to be exclusive and define a 'single best way' to polarise all the efforts of national governments and European decision makers. The complexity and apparent *lack of technological determinism* is actually good news for the EU. For instance, the recurring weaknesses in the production of ICTs does not mean that Europe is unable to grow faster and improve productivity, since these may be compensated for by potential strengths in the use of ICTs and the implementation of a knowledge-based economy (Cohen and Debonneuil, 2000). But, of course, decision makers have to be fully conscious of the related opportunity and take the needed structural measures.

Basically, each emerging growth regime calls for a precise institutional architecture, even though some reforms may enhance simultaneously the likelihood of several of them (Table 5.4). Conversely, given the precise economic, social and political configuration of each country, the chances of implementation, maturation and success for one regime may be greater than for another. For instance, Japan may still be leading in terms of modernisation of the post-Second World War mass consumption paradigm (Fujimoto, 2001), whereas the equivalent of industrial districts, such as Silicon Valley in the US, might be crucial for ICTs (Aoki, 2001) and competition-led growth still a characteristic of small open economies.

According to this diagnosis, the task of the European initiative is to *promote equally* the emergence of these diverse regimes, that may jointly contribute to European competitiveness and performance, while preserving the social cohesion that is characteristic of the long-run historical process of nation state formation.

3. SOME GENERAL PRINCIPLES FOR NATIONAL AND EUROPEAN POLICIES

Besides optimisation of the *short-run policy mix*, one major task for the public authorities is to create incentive constraints that push economic actors to innovate in the direction of one or another of the alternative *growth regimes*. Four major hints are to be discussed.

3.1. Convert the diversity of social systems of innovation into a strength

The frequent comparison of the US trajectory with the European integrating process leads to consider that Euroland is far from being an optimal monetary zone and therefore that the policy of the European Central Bank is bound to experience some problems (Gros *et al.*, 2000). Growth theorists add that the irrevocable fixing of internal exchange rates will foster a deepening of specialisation across countries and regions (Krugman, 1990); consequently, asymmetric shocks could be more frequent than initially expected.

Recent research on *European economic geography* suggests a more balanced view. Only a few sectors exhibit increasing returns to scale in the production of homogeneous goods, since the majority of them search for product differentiation by quality ladders and innovation in order to cope with extended competition (Maurel, 1999). Therefore, intra-European specialisation according to quality differentials could continue to prevail in the next decades, possibly with the exception of some sectors such as finance.

Table 5.4　Alternative emerging growth regimes and the redesign of institutional for

Institutional forms

Growth regimes	Wage labour nexus	Form of competition	Monetary regime
Toyotism	Exchange of employment stability against work malleability.	By quality and product differentiation of mass production.	Active Central Bank, promoting growth.
Service led	Strong heterogeneity/ inequality across industries.	Local oligopolistic for traditional services.	Trade-off between inflation and unemployment.
Information/ communication technologies led (ICT)	Dualism according to the ability to master ICT.	Linked to a dominant position in ICT	Role of risk capital and credit.
Knowledge-based economy (KBE)	Dualism according to schooling and cognitive skills.	Governed by the speed of innovation.	Credit, finance and even monetary policy pulled by innovation.
Competition led	External market flexibility and competitive wage.	Privatisation, deregulation, liberalisation.	Stability of monetary policy.
Export led	Frequently competitive, but exceptions.	By price and/or by quality.	Targeted towards price and exchange rate stability.
Finance led	Employment flexibility, profit sharing and pension funds.	Mainly on financial markets, but trend towards oligopoly.	Prevent the emergence of financial bubbles.

Thus, *the large diversity* of production and innovation systems could be preserved, even though of course some transformation and structural adjustments will take place during the next decades (Amable, Barré and Boyer, 1997a). This could be a strength since it would make it possible to reap jointly the benefits of contrasted social innovation systems (SSI): market based in the UK, social democrat for Sweden, Finland and Austria, as well as public institutions-based SSI in France, the Netherlands and Germany (Table 5.5). But, of course, national and European science and technology policies have to take into account the variety in the design of

State/society relations	Insertion into the international regime	Coherence and dynamic of the growth regime	Typical case
Developmentist state	Phase of export-led growth	The follower of typical Fordist growth regime	Japan until 1990
Strictly limited state, promoting flexibility.	Internalisation of modern business services.	Extensive growth with rising inequality.	US (1980s)
Building of public infrastructure for ICT.	New international division of labour according to the mastering of ICT.	Difficult to achieve for lagging countries.	Silicon Valley (since mid 1980s)
Schumpeterian welfare state	New international division of labour according to KBE	Difficult to implement in countries, with few academic resources.	US (1990s)
Pro-active and market enhancing state	Wider opening to international trade, investment, finance.	Risk of over-capacity and deflation.	Most OECD countries (since 1985)
New mercantilist strategy.	Key institutional form.	Strong dependency from external disturbances.	East Asian NICs (before 1997)
Under scrutiny by financial markets: search for credibility.	Trend towards global finance.	Risk of systemic financial instability.	US and UK (1990s)

adequate subsidies, incentives and statistical indexes of performance of SSI, and it is not necessarily an easy task.

3.2. Tailor economic policy to each national configuration given the precise externalities typical of a knowledge-based economy

With the process of globalisation, the detection of *best practices* has been a leading goal of firms and organisations. This benchmarking has been extended to public policy, for example by the Luxembourg summit on

Table 5.5 Three social systems of innovation are coexisting within EU

A previous work (Amable, Barré, Boyer, 1997a: 42–3) proposed a theoretical analysis about the existence of alternative social systems of innovation, gathered a series of statistical indexes in order to capture the major features of twelve OECD countries, showed the clustering of data around four configurations: the social-democrat model (Sweden, Norway, Finland); the meso-corporatist model (Japan); the market-based model (USA, UK, Canada and Australia); and the public-institutions-based model (France, Germany, Italy and the Netherlands).

It is interesting to notice that within the fifteen countries belonging to the European Union, three distinct social systems of innovation are observed.

The market-based model
The US, UK, Canada and Australia form this group. Competitiveness is low, unemployment and growth are at average levels. Other characteristics include difficulties in implementing post-Fordist productive principles; strong industrial specialisation in aerospace and chemicals; important military R&D; high level of publications/GDP but low levels of patents/GDP; flexibility of the labour market; and high level of expenditures in education but weak training. The financial system is sophisticated, and venture capital available.

The social-democrat model
Sweden, Norway and Finland constitute this group. Competitiveness is average, unemployment (during the 1980s) is low and growth is moderate. Foreign-based subsidiaries of domestic firms are important for their laboratories. Firms have adopted most of the principles of the 'new production model'. Industrial specialisation is strong in resource-intensive activities. Public research plays an important role, scientific specialisation is clear in medicine and biomedical research. Mobility and skills of the labour force are rather high. Education expenditure is high. The financial system is not very sophisticated, no venture capital is available and the cost of capital is rather high, but the ability to raise funds on international markets is good.

The public-institutions-based model
France, Italy, Germany and the Netherlands form this group. Competitiveness is rather high, unemployment (in the early 1990s) is high and growth is slow. Income distribution is favourable to the poorest 20 per cent of the population. Adaptation to the new productive model is average. Otherwise, this group is characterised by a certain number of specialisation characteristics. Scientific and industrial specialisation is marked in pharmaceuticals, aerospace and chemicals. Low level of scientific publications/GDP. External flexibility and mobility of the labour force is low. The financial system is relatively unsophisticated. Public expenditure in education is relatively high, but total education expenditure is moderate.

As we see here, the scientific profile does not mechanically determine technological specialisation, any more than technological specialisation mechanically deter-

Table 5.5   (cont.)

mines macroeconomic performance. It is neither necessary nor sufficient to have a dominant position in high tech to enjoy low levels of unemployment and high growth. Institutions that organise the innovation process are determinant: there is no ideal configuration to organise the relationships between science, technology and the economy.

employment policies (European Council, 1999b), or, more recently to technology policies (OECD, 1999). But it has to be remembered that a successful specific device may not succeed when inserted into a totally different institutional architecture: this is the central message of recent research about institutional complementarity (Aoki, 2001) and the hierarchy of institutional forms (Boyer and Saillard, 2001).

This hint is globally confirmed for innovation policies. All of them look at internalising the externalities associated with the very process of innovation, but, given the large variety of mechanisms involved, the exact economic policy tools may differ drastically. Just to give an example, enhancing human capital formation calls for different incentives than those required for tangible capital (Table 5.6). Furthermore, the permanent development of skills and competence may be the common outcome of *contrasted institutional settings*: an extended democratic educational system (Japan), implementation of a dual training system (Germany) or alternatively an intense retraining of workers hit by the obsolescence of their competence (Sweden) (Boyer and Durand, 1997: 53).

Thus, learning from others, adapting, tinkering and innovating might be the more valuable features of any relevant international benchmarking (Chapter 6).

3.3.   Organisational redesign: as important as tangible investment or RD!

During the early 1990s, Europe suffered from a low investment rate and the progressive recovery of profitability was counterbalanced by high real interest rates until the launching of the euro and the subsequent decline in the interest rate. Clearly, the unemployment rate seems inversely correlated to the investment share in GDP (European Commission, 1999: 10). One could expect that the wage moderation accepted by unions and workers, associated with the stable macroeconomic environment created by the rules set by the Amsterdam Treaty in terms of monetary policy, should now lead to recovery in investment, growth and after a while in employment. Such a

Table 5.6   Toward a knowledge-based economy: the need for new policies in order to cope with the related externalities

Nature of externalities	Intensity of impact	Robustness of estimates	Economic policy tools
Tangible capital	Strong	High	• Low taxation of capital, efficient financial market, stable macroeconomic environment
Intangible capital	Variable across industries (high in science-related industries)	Rather large	• Intellectual property rights • Interconnectedness of academia/research-industry • Tax and credit neutrality with respect to the choice between tangible and intangible capital
Knowledge and ideas	Assumed to be quite important for the future	Major difficulties, little robust econometric evidence	• Capital and labour mobility • High priority to education and training • Subsidy to networking and joint venture • Openness of national research to world community
Human capital	Important	Rather well documented	• Subsidy to education and vocational training • Definition of tracks and diploma • Easy access to credit by students
Learning by doing	Significant	Not so much statistical evidence, but some monographic evidence	• Employment stability but need for the stimulus by competition
using	Seemingly increasing	Little clear evidence	• Exposure of the population to new and generic technologies

Localised externalities (industrial districts, science parks, . . .)	Potentially very important (Detroit, Silicon Valley, Route 128)	Uncertain, but some monographic evidence	• Incentive to science parks • Quality of local infrastructures • Closeness of the links between education, research and entrepreneurship
Quality ladders	Positive upon firms profit	Indirect statistical estimates	• Polyvalence and high skills of workers • Demanding collective norms (minimum wage, environment)
Public infrastructures	From strong to vanishing	Rather problematic	• Detect the nature of infrastructures actually limiting growth • Redesign public budgets accordingly.
Forms of competition	Usually perceived as important, but ambiguous impact	In many high-tech sectors positive (telecommunications)	• Tune taxes in order to foster the optimum innovation intensity for the society

A. Investing still matters

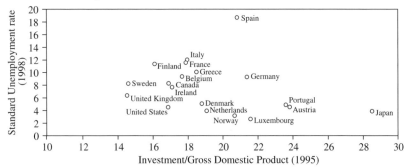

B. The Role of Computer Literacy . . .

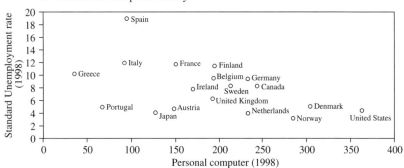

C. . . . And of Internet literacy

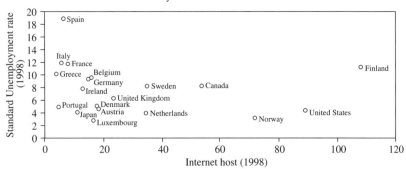

Figure 5.4 What strategy against European unemployment?

trend can be observed since 1998 for many European countries (Economie Européenne, 2000).

Nevertheless it would be too optimistic to consider that a good policy mix in the short run, on one side, and favourable demographic trends in the long run, on the other, are sufficient conditions for the progressive phasing out of mass unemployment. If the investment is made along wrong and now obsolete productive paradigms, the rate of return may even be negative with a totally adverse long-run impact on employment. *The quality of the decision process*, the relevance of the managerial tools of the firm and more generally the ability to implement the complete set of procedures that allows competitiveness in the contemporary world, all these factors are necessary to achieve positive effects from investment (Boyer, 1995). Similarly, the content of the investment is important and closely related to the quality of the internal and external information systems of the firm. This shows up from international comparisons: for instance, computer literacy and to a lesser extent access to the internet seem to play a role in reducing unemployment on top of the investment rate (Figure 5.4).

But, of course, it is a difficult task for public authorities to fine-tune intervention in accordance with such objectives, since the quality of internal managerial routines is not easy to assess by external auditors.

3.4.   Networking: the buzz word of the early twenty-first century?

The shift from the *linear model* – that used to convert basic science into economic performance via technological innovation – to an *interactive conception* – combining the synergy between innovation, market and production – has a definite impact upon innovation policies. The large public programmes, mission oriented, are largely replaced by more numerous but smaller projects with flexible economic objectives, associating all the relevant actors implied by the success of innovation.

Similarly, the large vertically integrated firm is no longer efficient in organising research and development if the competence belonging to various sectors and entities is to be combined. More generally, *horizontal as well as vertical networking* seems to be a key organisational structure in the era of ICTs, especially given the major uncertainties linked to the internationalisation of production and research.

Recurrent evidence shows that technological cooperation has a definite and large impact upon the success of innovation (European Community, 1997). But the question is the level at which such a networking should operate: in the case of ICTs, European joint ventures seem to have been less successful than cooperation with the American or Asian partners (Chapter 2). But a precise analysis of the related externalities (see Table 5.6 supra) suggests that the relevant level may vary drastically from the local community to the entire world, with a lot of intermediate levels including the nation or the region. It should be essential for European policy to carefully select the innovation sectors that are to be managed at the continental level. Even so, networking should be privileged.

4.   ADAPTING AND REFORMING ECONOMIC INSTITUTIONS

So far so good, since only the theoretical analysis is concerned. The real trick is to wisely implement such principles in a timely way. Then the most difficult problem is the respective role of regional, national authorities and finally the community level. Let us mention only two polar cases.

- If adequate negotiation and coordination mechanisms could be implemented at *the European level*, maybe a modernised configuration of Fordist mass production could be engineered. All the properties of the growth regimes of the 'Golden Age' would be translated from the national to the EU level. Of course, this is being quite optimistic about the relevance of this productive paradigm, in view of the fact that Fordism is more and more perceived as obsolete.
- By contrast, *the subsidiarity principle* may be invoked in order to delegate to national or regional governments the task to decide and implement the structural reforms on labour markets, welfare systems, public spending and taxation. For instance, the negotiation of national pacts or a complete decentralisation of wage negotiation define alternative strategies that deliver interesting results in terms of employment.

4.1.   At what level should externalities be internalised?

The answer is again closely related to the issue about the precise nature of externalities at work and the ability to implement coordination procedures in order to internalise them and get a more satisfactory outcome for the EU as a whole. The existing European Treaties define general principles about this distribution of responsibilities, but it is not totally evident that they are fully coherent both from a political and economic point of view (Table 5.7).

Let us review the different components of European and national economic policies:

Table 5.7   At what level should each economic institution and economic policy component operate?

Domain	Nature of the externalities/ coordination problems	Institutional arrangements
Monetary policy	Monetary stability enhances the single market and financial integration	*European Central Bank*, as a federal entity
Budgetary policy	Spill over from one country to another of any national deficit or surplus	A *common rule* for each country: the stability and growth pact embedded in the Amsterdam Treaty.
Tax policy	Regime competition for the more mobile factors (capital, professionals)	*Ad hoc coordination* among Ministries of Finance in response to emerging problems.
Policy mix	Strategic interaction between monetary and budgetary policy	Coordination within an *informal group* (Eurogroup)
Wage/ employment	● Limited labour mobility across borders ● At the structural level, spill-over from managerial strategies (restructuring) to the financing of welfare (early retirement)	A large variety across countries: ● *Negotiation* of national pacts (Netherlands, Spain, Italy, Finland). ● *Decentralisation* and more heterogeneity across various forms of labour contracts (UK, France).
Competition	Possible distortion by national policy of fair competition on the single market.	Monitoring of competition by a *European body*, but national enforcement.
Innovation	The externalities may be operating at: ● The world level (science, pharmaceutical, software). ● The European level (transport infrastructures, coordination of research centres and firms). ● The national level (legacy of each system of innovations . . .) ● The regional level (e.g., industrial districts, science parks)	The related *institutions* differ: ● Open science and uniform/ universal patenting. ● European programmes fostering networking (subsidies, credit). ● National science and technology policies. ● Regional policies
Insertion into the world economy	● Strategic use of competitive devaluation. ● Strategic and new interaction between exchange rate policy of the EU, the US and Japan.	● A single exchange rate for all members of Eurogroup. ● Rules governing exchange rate policy for other countries ● *Single representative* in international negotiations about trade. ● Shared responsibility for finance between ECB and head of European Council.

- *Monetary policy and exchange rate* management are irreversibly a matter of common concern, the task being given mainly to the European Central Bank with some role to the European Council about the strategic choice of an exchange regime.
- By contrast, *budgetary policies*, that exert similar externalities, are kept under the control of national authorities, but the Stability and Growth Pact imposes a limit to public deficits as well as sanctions (Conseil d'Analyse Économique, 1998). This discrepancy has been playing a role in the muddling through observed in the European policy mix after January 1999, as explained more fully in the next section.
- *Competition policy principles* are common to all member states, but implementation and control remain national. In the long term this might trigger the constitution of a fair trade and competition agency, US style. This would mean a large breakthrough by comparison with the previous juxtaposition of contrasted national styles for governing competition (Dumez and Jeunemaître 1996). Note that sanitary and security standards for food and some manufactured goods has too to be managed at the same level as the single market. This was recognised by the conclusion of the Nice European summit.
- *Innovation* is mixing a large degree of subsidiarity along with European programmes aimed at building transborder networks in the domain of applied research or of academic activity. Maybe with the evolution toward a KBE, the mobility of scientists and professionals across Europe should be promoted more actively.

A theory of such a complex socioeconomic system remains to be elaborated and the task is overwhelmingly difficult; it will take some time and a lot of learning and experimenting. Thus, meanwhile some pragmatic rules have to be invented and discussed in order to initiate a quite pragmatic process, with largely unintended configurations one or two decades later (Boyer, 1999b).

4.2. Mixing short-term management with long-run structural policies

Decision makers are facing *two dilemmas*.

- The first one relates to the fact that *the time required for any institutional reform* to deliver the expected (or unexpected) results is far longer than the time of polity, that itself is rather long with respect to the speed of finance. The scrutiny by financial markets of any government decision is an incentive for public authorities to focus upon short-term issues. But in the long run, everybody is now convinced that structural reforms are required, but they may take ten or fifteen years to deliver all their (hopefully) positive outcomes. The reforms of the Dutch welfare system and the labour contracts are a good example of such *a long lag* between the negotiation of a social pact and the resulting decline in unemployment.

- The *second dilemma* concerns the links between *the timing of structural reforms and business cycles*. When the economy is booming, governments are happy to distribute the related surplus and do not perceive any need for painful structural reforms, since they are enticed to consider that 'the crisis is over!'. They could do reforms but they are not induced to do so. Conversely, when there is a major crisis, the need for structural reforms becomes self evident, but the cost is so huge and the lag so important that they are postponed to better days. Even if they are done it is too late to overcome the current crisis.

Thus the real trick is to conduct a 'wise' short-run economic policy, while sequencing the structural reforms that will be operative several years later. Again, the good prospect of European growth for the first years of the millennium enlightens the temptation to postpone quite necessary reforms.

4.3.   The Euro: working out a new and satisfactory policy mix

A fraction of European unemployment in the 1990s could be attributed to an unsatisfactory policy mix, caused, among others, by the absence of coordination of national budgetary policies in response to the German reunification (Conseil d'Analyse Économique, 1998). Therefore, any improvement in the dialogue between the partners of economic and social policy is welcome. *The communication* between the ECB and the Ministers of Finance has been improved by the creation of the informal Eurogroup, besides the official Ecofin Council. *The learning process* that has thus started may initiate an interesting scenario in order to overcome some of the ambiguities or weaknesses of the Amsterdam Treaty (European Commission, 1999a; Boyer, 1999d; Jacquet and Pisani-Ferry, 2000; Boyer and Dehove, 2001)).

Thus, the implementation of the euro is part of a deeper move in the conceptions and tools of contemporary economic policies. Back in the 1960s, a highly centralised management of the policy mix was organised in most countries by the national Ministries of Finance that had clear control over the Central Bank. Information originated mainly from the state and was inserted into macro modelling and annual forecasting exercises. By contrast, in the 1990s, the main players became formally more independent but had to exchange information about their strategy: the actual policy mix is the outcome of this complex process (Table 5.8). In a sense, the creation of the ECB is a step in that direction, with the need to organise *a forum for coordinating strategies* that have now to operate both at the European level and at the national level.

Table 5.8   The new style for economic policy: a condition for optimising the European policy mix

	The top–down approach	The relational approach
Key actors	National Ministry of Finance	Independent European Central Bank
Centralisation/ decentralisation of decisions	Strong centralisation	Shared responsibility of the policy mix
Source of efficiency	Coherence of a centralised program	Credibility, reputation compatibility with private actors' expectations
Information flow	From public authorities to private agents via public statements	Informal cross exchanges among separate public bodies and major private players
Method to make decision	Mainly centralised tools with little concerted action	Learning from interactions between actors with different objectives
Tools	Macroeconomic modelling with static or adaptative expectations	Rational expectations models or at least forward-looking reasoning
Emblematic example	French Ministry of Finance	Eurogroup
Periods	1960s and 1970s	1990s to early next century

Source: Freely inspired from EC/DGE 'Reinforcement of mechanisms for economic policy coordination', 28 July, 1999.

The decision of the *Cologne European Council* to implement a macroeconomic dialogue may mean a new stage in this process of *mutual information and coordination*. By inviting social partners to discuss wage and employment issues, some emerging conflicts could be cured at an early stage and this could significantly improve the performance of the EU in terms of job creation (Maurice, 1999). Price stability would be largely warranted by collective negotiations delivering near stability of unit labour costs, so that the ECB could deal with symmetric shocks affecting European growth, while delivering monetary stability.

Nevertheless, the European policy mix is facing *a major challenge*. On one hand, the Amsterdam Treaty formalises a clear asymmetry between the Central Bank, which is given both clear autonomy and precise objectives, governed, first, by price stability and, second, by a contribution to general economic policy and the national governments which have to comply with the Stability Growth Pact that puts an institutional constraint upon the autonomy of each national budgetary policy, precisely to preserve the credibility of a low inflation monetary policy. But, on the other hand, the ECB policy is itself under the close scrutiny of international financial markets, that assess, on a permanent basis and in real time, the viability of the current policy mix. Consequently, the European monetary policy is itself disciplined by the rules and evolution of the international economy. This feature makes the process operating within Eurogroup specially important and by extension the process implemented by the Cologne European Council.

4.4. Same monetary policy, but different unemployment performance

If organisational choices matter for firms' performances, similarly institutional reforms matter for employment creation, for any given policy mix. Contrary to the idea conveyed by typical Keynesian theory, much evidence suggests that European unemployment is not totally caused by pure coordination problems between monetary and budgetary authorities. The nature of collective negotiations, the incentives brought by the welfare and tax systems, the degree of reactivity of hours, wages, bonus and employment to various shocks do matter for macroeconomic performance.

- Among the *medium-sized European countries*, the pattern for unemployment is quite different between continental Europe and the UK, especially since early 1993 (Figure 5.5a). These differences can be attributed to the degree of decentralisation and privatisation of public services, or slimming down of welfare compensation. This is the origin of the strategy that tries to develop new security and social rights, benefiting to the 'employability' of workers, according to the ideas developed by the 'third way' (Giddens 1998). By contrast the strategy 'wait and see' has proved to be quite detrimental.
- *Small open economies* display, on average, far better results since they benefited, at the end of the 1990s, from a quasi full employment (Figure 5.5b). This can be explained by the fact that since the 1950s, social partners have taken into account the need for competitiveness

A. Medium-sized European countries: a premium to social deregulation?

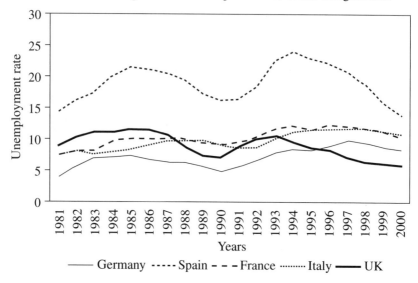

B. Small open economies: the primacy of new social pacts?

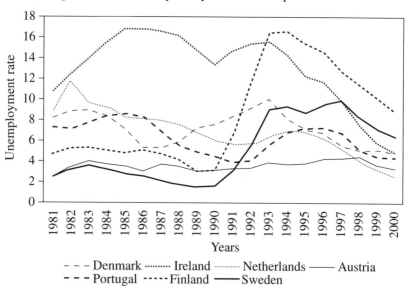

Figure 5.5 European Union: the same macroeconomic environment but contrasted unemployment rate evolutions

and expressed a major concern for full employment. Nevertheless, the discrepancy between the inherited welfare and tax system and the new pressures associated with internationalisation triggered the need for major reforms in the mid 1980s. When *national social pacts* have been struck, the outcome seems to have been quite favourable indeed. The Netherlands is a good example to this alternative to the *Third way*.

Consequently, in the long run *social bargaining matters*, along with fine-tuning the policy mix in the short run. This gives a lot of meaning to the subsidiarity principle and the significance of responsibility to social partners. The Luxembourg process may help in pointing out the diversity of institutional arrangements across European countries and give some hints for reforming the legal and conventional procedures that have proved to be detrimental to job creation and social cohesion (European Commission, 1999c, 1999e). But mere importation and imitation are generally not sufficient, since national industrial relation systems are quite idiosyncratic and call for adaptation of any one best way, the search for functional equivalents and even radical innovation. This is observed both at the level of sectors (Boyer, Charron, Jürgens and Tolliday, eds, 1998) and of innovation systems (Chapter 6).

4.5. Adjusting wage and employment negotiations to the common monetary policy

These contrasted performances may represent an incentive for lagging countries to foster significant reforms in order to be sure that the unemployment levels will not be the disciplining device for curbing labour costs and contributing to monetary stability. The process towards EMU and the launching of the euro have clearly triggered major concern for this issue from social partners (Maurice, 1999).

- On one hand, the task may seem easy since at least *eight solutions* are available, with contrasted features and outcomes in terms of dynamic efficiency and social justice. Social partners could pick their preferred solutions from among this rather large menu (Table 5.9). However it seems that fully integrated collective bargaining at the European level is, for the time being, out of reach, as far as quantitative issues are concerned (Kim, 1999). Clearly it is quite difficult to make compatible a set of national wage demands, given the remaining large heterogeneity of productivity levels within the same sectors across Europe.

Table 5.9   What reform of industrial relations in order to cope with European

Configurations/results	Strengths
1. Euro-corporatism Collective bargaining at a European level according to the European Central Bank's monetary policy	● Entirely compatible with the new monetary syst ● The German model is transposed to all of Euro
2. Xeno-corporatism Collective bargaining within the European multinationals	● Emergence of a pan-European type of solidarit between employees ● Extension of the 1994 EU directives concerning the circulation of information within transnational companies.
3. Interlocking industrial relations Multipartite bargaining and interlocking systems of structuring industrial relations 　At the European level: depends on price expectations 　At the national level: depends on productivity and solidarity 　At the company or industry level: profit-sharing	● The country reconciles its macroeconomic constraints with the situations in its various sectors and firms ● The system is coherent with the way in which power is distributed to collective actors – not all of whom have the same hierarchical level
4. Hierarchisation of industrial relations The German tripartite mechanism is the first to set wages according to the European Central Bank's monetary policy. Afterwards, tripartite institutions in the other European countries negotiate their own trade-off between wages and employment	● To preserve the differentiation of national practices, the supervisory institutional mechanisms must become flexible ● The aforementioned asymmetry (between Germany and the other European countries) is perpetuated, but monetary policy can be transposed into the wage-setting realm
5. National neo-corporatism The national pacts adopted by the firms and employees in each country in light of the policies of the European Central Bank.	● The coherence between existing laws and institutions facilitates compromise of this kind ● An opportunity to negotiate reforms that can improve the job market
6. Meso-corporatism The whole sector becomes involved in collective bargaining at a European level	● The variability of each sector's competitiveness i taken into account ● Forces that support this kind of negotiation are emerging inside Europe
7. Micro-corporatism Negotiations are only held at the company level	● Attempt to improve the trade-off between wages and employment ● Corresponds with the 1980s and 1990s trends towards decentralised wage bargaining
8. Labour markets Competitive labour markets are set up	● Full employment (at least in theory) ● No need to coordinate collective structures ● Corresponds to the US model – a 'job-making machine'

onetary policy?

rtcomings	Likelihood
:w actors are capable of negotiating within this sort ` framework side of Europe, there are many different conceptions ` how the Union should turn out	From likely to somewhat improbable
mited applicability, may lead to the further •terogenisation between employment contracts in ıtional vs. international firms ɪpsets the balance of power between the tripartite stitutions – capital can migrate more readily than ɪployees	Emerging, but of limited scope
ɔmplexity of an approach that requires separate but ɪterlocking negotiations ncertainty as to the level at which negotiations should ɪ carried out. Plus, who should be the catalyst – local ˈms or the European Union?	Conceivable – might even be probable
fragile mechanism that can be destabilised by ɔmestic German political interactions (cf. the 1999 ɪge adjustment negotiations) . theory, the Euro is supposed to establish a •mpletely pro-rata symmetry between the various ɪember countries, reflecting their weighting of the ɪacroeconomic aggregates ɛrtain countries lack any tripartite institutions that ɪe willing to adopt this strategy	Very likely (*de facto* coordination between Germany, Holland, Belgium and Austria) but problematic in countries where negotiations are totally decentralised (already the case in the UK, soon to be in France)
ɔmpanies may prefer complete decentralisation to ɪis new version of neo-corporatism ɪgh unemployment can weaken unions' bargaining ɔsition	Very likely (already operational in Italy, Spain, Holland and Finland)
ɪfferent sectors have very heterogeneous pay systems ˈhat to do with protected national industries?	Possible in a small number of sectors (e.g., automobiles, chemicals, freight)
employees are not particularly mobile, wage ɪfferentials can rise . certain configurations, unemployment coexists with ɪflationary tendencies (UK in the 1980s)	Very likely in countries where very few negotiations are centralised
ɪndermined by modern microeconomic theories •ncerning the prerequisites for full employment ˈrong unions can keep this from happening ˈhat to do if the free market loses popularity •cause it engenders inequalities (i.e. France)?	Not to be excluded, but not inevitable

- On the other hand, in the absence of well-organised business associations and workers unions or if, while existing, they are prone to conflict and the adoption of 'loss–loss strategies', most of these cooperative solutions are out of reach (Boyer, 1998b). The only path available is then large social deregulation, which would be a threat to domestic social and political stability and could mean the end of the ideal of a social Europe (Maurice, 1999; Freyssinet, 1999). It has even been argued that the market cannot elicit the requested commitment from workers by pure market mechanisms, without provoking in the long run major instability (Bowles and Boyer, 1990). This argument is still more relevant in the emerging *new economy*. (European Commission, 1899c.)

With respect to the Luxembourg procedure that launched the benchmarking of employment policies, a lot of initiatives have to be delegated to national or regional actors in order to select what innovations could possibly be imported and adapted to the local context. Again the subsidiarity principle is quite essential for the future of the EU.

4.6. Redesigning all legal and tax incentives in order to promote the adoption of new growth regimes

All experts, economists (including the Keynesians!) and of course the politicians should take seriously the warning from Keynes. Most of us are consciously or not following the ideas and theories of economists of the past, who themselves were analysing a now largely obsolete economic system or vanishing 'regulation mode'. The contemporary institutional systems have been built from the sedimentation of a series of compromises and legislation over one or two centuries. The task is now to *scrutinise this complex architecture* in order to detect possible shifts in these compromises or, in some cases, to propose brand-new arrangements. But the real difficulty is then: how to build a self-sustaining political and social majority in favour of such reforms? Just to mention some items at the top of this agenda:

- Given the issue of the ageing European population, it is important to shift from *unemployment policies* that aim at reduction of activity rates for old workers towards a more *active policy of job creation* (European Commission, 1999a; European Council, 1999a and b). The facility of early retirement should be phased out, the more so the more dynamic the European economic recovery and the more likely the emergence of labour scarcity for some professions (Taddei, 1999).
- A plan for training the less-privileged members of European societies would be welcome for many reasons. From an economic point of

view, whatever growth regime prevails during the next decade (ICTs or KBE), the access to *general education* is more essential than ever. The trend towards a learning society is a common feature of the likely evolution. Similarly, the changing productive paradigm may hurt the practice of on-the-job training and call for a renewed interest in *vocational training systems*, promoting some transferability of skills (Caroli, 1995). This is specially important given the trend towards large heterogeneity of labour contracts according to level of competence and transferability (Beffa, Boyer and Touffut, 1999). Some workers leave the educational system without any clear competence and are unable to acquire any professional expertise, given the large mobility they experience from one low-skill job to another. The objective of the reduction in social inequality is thus more essential than ever. Last but not least, modern democratic societies need informed and learned citizens.

- The *welfare systems* that had been designed in order to fight against the social inequalities observed after the Second World War have in a sense succeeded. For instance, in most European countries, retired people are no longer poorer than the rest of the population. But, in some cases, various components of the welfare system have become counterproductive (concerning for instance the rise of labour costs due to the financing of welfare) and still more these systems totally neglect the new sources of inequalities (e.g., youth unemployment, social exclusion, unequal access to education, single-parent families, emerging urban ghettos). This is a major topic for social security reforms (Esping-Andersen, 1999, and Chapter 3).

- The reform of the *tax system* is facing another dilemma. On the one hand, a Schumpeterian conception would imply a shift of taxation from innovators and entrepreneurs to the new rentiers who have prospered during the last fifteen years. But, in turn, this could trigger wide income differentials when, for instance, stock options dramatically inflate the revenue of top managers by comparison with rank-and-file workers. On the other hand, the social heterogeneity brought by the diffusion of market-led capitalist institutions puts a new emphasis on the agenda put forward by John Maynard Keynes during the 1930s, even though the regulation modes and international regime are not at all the same. But then governments may fear a brain drain of the most talented individuals, a loss of dynamic efficiency and finally less growth and less employment. European contemporary social democrats are muddling through this dilemma.

- Last but not least, the EU should contemplate extending *financial supervision* at the community level in order to prevent the emergence of speculative bubbles that could be quite detrimental to growth stability in the medium to long term. For the time being, the issue is hotly debated among experts. On the one hand, moral-hazard arguments would attribute the role of supervision over their domestic banking and financial systems to national authorities and confirm the clauses of the Amsterdam Treaty. But, on the other hand, what about the globalisation of finance and the spill-over of the systemic risk from one country to another? (Conseil d'Analyse Économique, 1998; Aglietta and De Boissieu, 1998; 1999). In the long run, will not the European financial system be integrated and call for a common supervision?

This is an invitation to consider what new domains should be attributed to European Community responsibility, thinking ahead to any emerging problems. But conversely, other areas of present intervention should be phased out.

4.7. Recompose the European budget around growth and social cohesion objectives

By constitutional design, the European budget cannot play any role in the formation of the policy mix, since no deficit, or surplus, is allowed, and no tax is levied directly at the European level. Therefore, the objective should be to optimise the contribution of the European budget to the long-term dynamism of the Old Continent, along with the implementation of some solidarity across national and regional borders. Such a process of European budget restructuring has taken place during the last decade, but it should continue. Here are some possible guidelines.

- *The common agricultural policy* is evolving from administrative price formation of selected agricultural products to income maintenance of some disadvantaged farmers who would eventually take charge of environment preservation in rural areas. Ecological concerns and solidarity with farmers would shape totally new public interventions.
- At the national level, subsidies should be shifted from mature industries to *sunrise industries*, as far as such a strategic move is allowed for by the political configuration that governs typically for each society. Furthermore, the numerous lobbies present around the European Commission seem to be more active in mature and declining industries than in sunrise and emerging industries.
- A new design of *regional and structural European funds* would be useful, in order to take into account that some targeted regions are

now developed, that new sources of inequalities may emerge from this new phase of European integration and finally that the admission of new members will increase heterogeneity and call for new solidarity principles.

- Finally, the *targeted socioeconomic research programmes* should resist the temptation to launch large mission-oriented public projects. It would be far better to propitiate the building of European networks, mobilising medium-sized and small firms as well as large corporations along with the academic world. To convert the diversity of social systems of innovation into a European asset would be a rewarding objective that would give a lot of adaptability to Europe, given the coexistence and dynamism of contrasted growth regimes among the EU.

But there is a major difficulty: how to actually implement such broad orientations? It is not an easy task. By chance, the recent processes launched by the various European Councils provide some tools.

5. HOW DO THESE IDEAS FIT INTO THE EUROPEAN AGENDA?

The context of the first years of the millennium is quite good, since many conditions governing the appeal and feasibility of such a programme are fulfilled. Nevertheless, the agenda is far from simple and many obstacles have to be overcome. These short remarks are quite preliminary and introductory.

5.1. Convert the current recovery into a long-term boom

It is important to stress again that short-term recovery does not mean the end of all the structural problems experienced by the member states in the EU. Quite the contrary, all the actors should be convinced that the current European institutional architecture is far from being totally coherent (Boyer and Dehove, 2001), that some reforms have to be made to the national industrial relations systems, the links between schools, universities, research centres and firms, to the financial organisation of welfare, not to mention to the tax system itself.

When, during the 1990s, European leaders used to look with some envy to the so-called American 'New Economy', they should have noticed that the remarkable performance of the 1990s had been reached after a series of structural reforms, starting back in the early 1970s and affecting nearly every aspect of social and economic life. Even so, the long-run viability of such a regime is still to be assessed (Boyer, 1999c). The problem is still more acute for Europe, since, for some if not all countries, major reforms are still to be made in order to consolidate the current optimistic expectations and growth perspectives. Only a well-ordered flow of reforms will transform the current recovery into a long-term growth regime.

5.2.   Broad economic policy guidelines: a mirror image of the new institutional architecture

Many policy tools have been created in order to coordinate the national reforms, and these initiatives may seem, to some distant observer, as highly *ad hoc* and probably quite difficult to implement. The present report brings a kind of optimism, but stresses the need to *simplify the complex architecture* of these various processes. Basically, the actual architecture of the relations between the monetary and exchange rate policy and the Broad Economic Policy Guidelines (BEPG) is a mirror image of the emerging architecture of the institutional forms after two decades of restructuring of the Fordist regime (Figure 5.6).

For instance, the redesign of the objectives and tools of public interventions, implied by the concern for credibility and the emergence of a lean and smart state, could be propitiated by the stability and convergence programmes. Initially conceived of as methods and principles for controlling by peer review excessive public deficit, they should evolve towards an analysis of the efficiency of various public spending strategies (Jacquet and Pisani-Ferry, 2000). Similarly, the Luxembourg process defines guidelines for making the wage labour–nexus more reactive to technical change, macroeconomic fluctuations, gender equity issues, and so on. The Cologne process is part of this evolution, since it recognises that a viable 'regulation mode' has to make the initially disconnected transformations of monetary policy, budgetary strategy, wage bargaining and competition compatible. The strategy adopted at the Lisbon summit builds upon the challenge addressed by the KBE in order to promote a series of reforms in all areas of economic and social activity. Therefore, the various European processes are in tune with the major structural transformations taking place both at the international and national levels: this feature creates some room for success.

5.3.   Making virtuous circles? A method for coordinating complex European procedures

Quite rightly, most recent reports from the European Commission and Council have recurrently pointed out the dangers associated with the multiplication of the European processes: they may have large administrative costs, trigger possible conflicts among them and may lack transparency. For

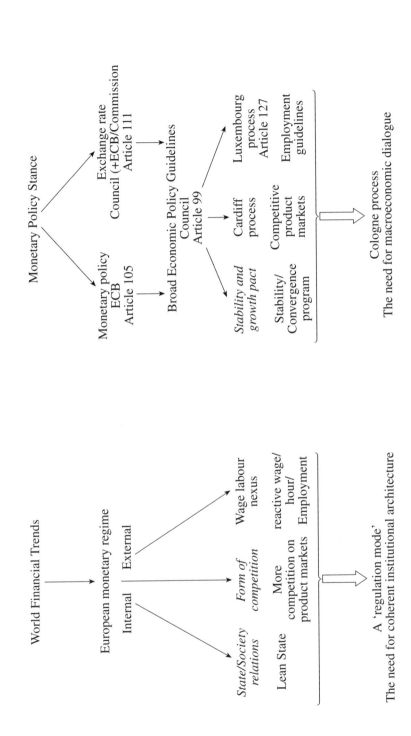

Figure 5.6 A good news: the new hierarchy among institutional forms is taken into account by the structure of European treaties and subsequent decisions

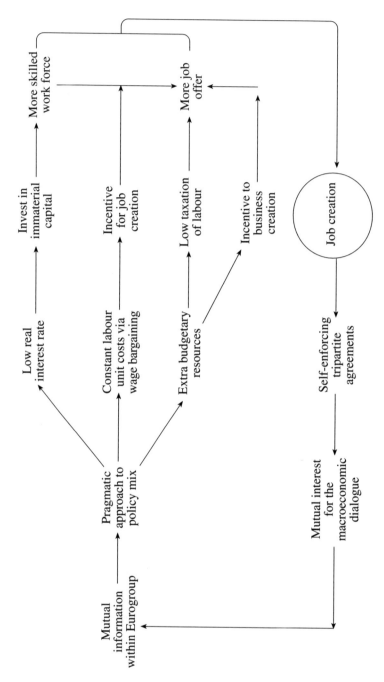

Figure 5.7 Strategy one: use the dividend of faster growth to lower the tax and remove welfare-related barriers to job creation and launch the macroeconomic dialogue

many domains, the various initiatives overlap (European Commission, 1999a; European Council, 1999a, b). Without any clear orientation, the whole architecture may well be ineffective and call for easy criticism about the excessive bureaucratic burden associated with Europe. The effectiveness and transparency of European procedures is quite important for the legitimacy of European integration, and the adhesion of citizens (Quermonne, 1999).

The present analysis provides one hint: why not consider that the unit of analysis for public intervention is not the precise domain (e.g., tax, welfare, industrial relations) but the *whole set of positive spill-over* that may define a coherent development mode or at least a complete set of mechanisms linking the various domains? This is no more than converting the analysis of the third section into a practical tool for assessing the coherence of a strategy. This could possibly define *new frontiers among the existing Cardiff, Luxembourg and Cologne processes*. Only four examples will be given.

- *Channelling the dividends of a relational policy mix* is a first and important option, since it relates the short-run to the medium- and long-run perspectives (Figure 5.7). A pragmatic approach to the policy mix, based upon mutual learning in a densely organised arena for the exchange of information, may be an ingredient for a steady boom in investment, directed towards material and immaterial components, thus generating a positive and increasing sum game. This issue is under the responsibility of ECB and national Ministries of Finance. Such a virtuous circle, relatively well defined, is at odds with the vicious circle that emerged during the early 1990s in Europe. But then, the reforms concerning competition policy, labour market organisation and public budget should be explicitly coordinated according to this specific virtuous circle (Table 5.10). Macroeconomists could be the architects of such a strategy, in response to the strategy adopted by the European Councils and debated within the European Parliament.
- *Organising the shift from ICTs to KBE* is another method for trying to initiate a virtuous circle of innovation, growth and job creation. This is one of the interpretations of the 'New Economy' and this strategy should be implemented by Ministers of Science, Technology, Education, of course in close connection with social partners (Figure 5.8). It is important to note that this mechanism is largely independent of the previous one and calls for different institutional reforms that might therefore be congruent with the first strategy (Table 5.11). In contrast with the previous virtuous circle, this one is specially appealing for private actors, entrepreneurs, employees, teachers, researchers and so on.

Table 5.10 The institutional setting for strategy one

	European level	National level	Decentralisation
Wage labour nexus	European collective agreements	National pacts	Decentralisation of wage formation
Form of competition	Possible contribution of the extension of the single market		
Monetary regime	Reaction to symmetric real shocks		
State/Society relations		Pro-labour reforms of the tax and welfare financing	
Insertion into the international regime			

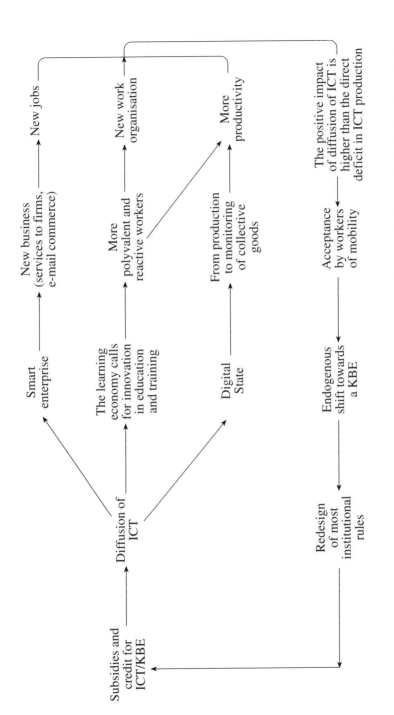

Figure 5.8 Strategy two: convert the information and communication technologies (ICT) into the basis for a knowledge-based economy (KBE)

Table 5.11 The institutional setting for strategy two

	European level	National level	Decentralisation
Wage labour nexus	Mobility of scientists and professionals	Ambitious programs for training and retraining workers to ICT and the KBE	
Form of competition	Facilitation of business creation, clear property rights for information/knowledge		
Monetary regime	Stable monetary rule with low real interest rate. Active financial trading in order to transfer capital from mature to sunrise industries		
State/Society relations		Provide the education, infrastructure and tax incentives for KBE	
Insertion into the international regime	Import of ICT against export of KBE		

- *Welfare reform-led virtuous circle* may well be a major trump for Europe. This might sound quite strange in an era when the burden of the European welfare is frequently considered as the major obstacle to European job creation. Clearly the idea is not to keep unchanged the complex welfare systems progressively built during the 'Golden Age' and amended afterwards, but to reform them in accordance with new social demands and macroeconomic opportunities and constraints. Better implementation of gender equality and the phasing out of early retirement could trigger a surge in the activity rate and generate new demand, all movements that would ease the financing of welfare systems (Figure 5.9). These would need to be redesigned in order to cope with the new forms of social exclusion, related for example to an inadequate initial general education and the inability to master modern technologies. This is a matter for Social Affairs Ministers and social partners. Again, the institutional reforms necessary for the implementation of such a mechanism are not necessarily contradictory to the previous ones (Table 5.12). The benefit of such an approach is to provide a clear rationale and various tools for coordinating a series of reforms and to check that they are more complementary than contradictory.

- *Coping with a finance-led regime* could define another alternative, specially relevant for the countries already specialised in financial intermediation and business-related services (Figure 5.10). The objective of the ECB should then be to incorporate the curbing of asset inflation; social partners should negotiate profit sharing and pension funds management, and the supply of welfare could probably be more or less significantly transformed by privatisation of many components, including retirement payments (Table 5.13). The logic of such a mechanism seems so strong, that, according to some extreme scenarios, it may enter into conflict with the three previous virtuous circles. Furthermore, it is not sure that the related regime is stable in the very long run. In any case, it is far from being widely diffused across Europe, especially if new members joined the EU. Nevertheless, this framework may help in conceptualising the issue brought by the implementation in Europe of pension funds. Can Europe find out some functional equivalents to the American system or should the whole institutional setting be redesigned? This is a major political choice, that has to be enlightened by more investigations.

These proposals raise another important issue: at what level should these virtuous circles operate?

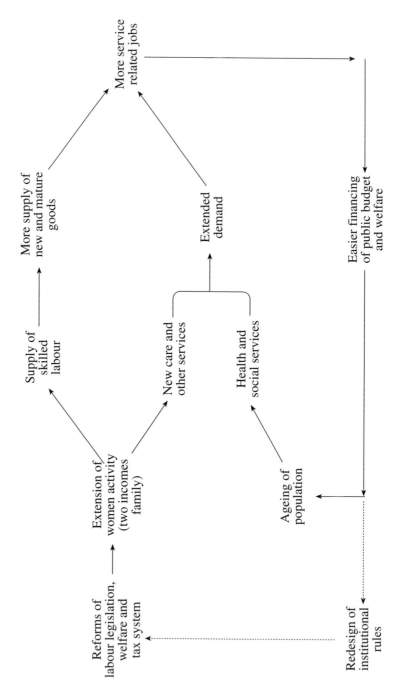

Figure 5.9 Strategy three: gender equality and responses to ageing as the source of a new service-led growth

Table 5.12   The institutional setting for strategy three

	European level	National level	Decentralisation
Wage labour nexus	Promotion of gender equality	• Life-long cycle of activity • Extension of retirement age	
Form of competition	Facilitation of new services	Tax reduction for the services sheltered from international competition	
Monetary regime			
State/society relations		Completely redesign welfare for a two-income family	
Insertion into the international regime	Relative autonomy of a welfare-based growth regime		

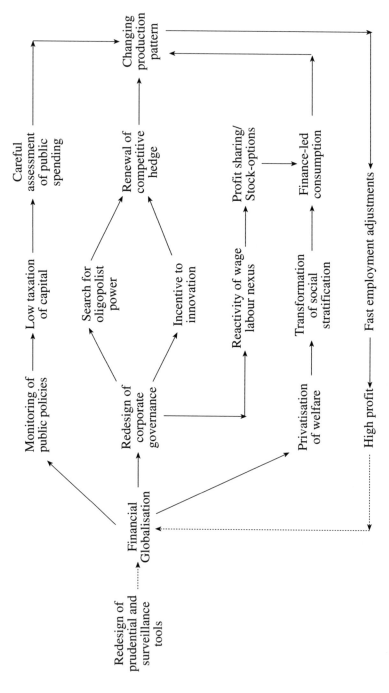

Figure 5.10 Strategy four: riding the financial globalisation

Table 5.13 *The institutional setting for strategy four*

	European level	National level	Local
Wage labour nexus	General guidelines about profit sharing/stakeholder society	Negotiation about the sharing of the dividends from finance	Highly differentiated formula across firms, regions
Form of competition	Enforcement of fair competition in spite of oligopolistic trends		Possible exceptions for local activity sheltered from international competition.
Monetary regime	Preventing asset inflation, as another objective of ECB	Sharpening of national enforcement of prudential ratios and control	
State/society relations	General guidelines	Privatisation of some components of welfare. Monitoring of the private production of public goods	
Insertion into the international regime	Role in the restructuring of the international architecture of finance		

5.4.   Putting the subsidiarity principle at work

It is clear from the previous developments that some parts of it operate at the European level but are necessarily linked to mechanisms operating at the national or even regional levels. It can be argued that some member states may follow one regime, whereas others adhere to a different one, while complying to the same European rules and general evolution. After all, it is precisely the meaning of the *subsidiarity* principle, that *a priori* grants some autonomy to each local authority in order to follow the common European principles and rules. Furthermore, these rules do result from the interaction of national governments and they thus link the various levels of governance: regional, national and European.

This is specially important for the long-run acceptability of European integration (Quermonne, 1999). It has to be simultaneously *politically legitimate and economically efficient*: the subsidiarity principle and the emergence of jointly decided rules of the game may provide these two requisites.

6.   CONCLUSION

During the 1990s, the sharp contrast between the US booming 'New Economy' and the ailing Old Continent has fed a lot of Europessimism and doubts about the ability of the European Union to work out a new and efficient growth regime. In the early 2000s, the scenery has somehow shifted and a more balanced view is emerging.

On the one hand, the euro has provided a new impetus to the Single Market, somehow protected Europe from the consequences of the 1997 Asian crisis and started a far-reaching experimental and learning process for both private and public actors. Concern for euro/dollar parity should not hide the more structural issues at stake. How to overcome the lack of budgetary federalism with open cooperation between national member states via the emergence of commonly agreed and nationally implemented rules about taxation of mobile factors, employment and social policy, innovation strategies and so on? This chapter has provided some analyses and preliminary orientations, that will not be summarised again.

The core idea is to realign most if not all European guides lines by reference to emerging virtuous circles. That may imply for instance converting ICT mania into steps towards a KBE, much more diversified and long lasting, or making a competitive asset out of a reformed but still well-developed European welfare state.

On the other hand, the related reforms are not that simple. First, the sedimentation of half a century of legislations and compromises makes the architecture of the European Union quite complex, even for the real experts, not to speak of the potential new members that are scheduled to join the EU. Second, the wave of deregulation of the 1980s and 1990s has created new externalities and sources of instability that spill over from one country to another. Each government in isolation has no longer the power to fix some emerging imbalances, but conversely the European Councils experience a lot of difficulties in agreeing upon new principles. The outcome of the Nice summit might be a good example of such a dilemma.

Last but not least, governments are reluctant to make structural and initially unpopular reforms when the economy is booming, but it is too late to make them when the crisis hits! Is the 1986–91 episode to be reiterated or will the 2000s be perceived as a turning point and the beginning of an epochal change? So many innovations and potential unbalances have been taking place during the 1990s that it is not totally unwise to bet for the latter.

REFERENCES

Aglietta, Michel (1998), 'Le capitalisme de demain', *Notes de la Fondation Saint Simon*, 101, November.

Aglietta, Michel and Christian De Boissieu (1998), 'Problèmes prudentiels', *Conseil d'Analyse Economique*, 5, Paris: La Documentation Française, pp. 49–59.

Aglietta, Michel and Christian De Boissieu (1999), 'Le prêteur international en dernier ressort', *Conseil d'Analyse Economique*, 18, Paris: La Documentation Française, pp. 97–127.

Amable, Bruno, Rémi Barré and Robert Boyer (1997a), *Les systèmes d'innovation à l'ère de la globalisation*, Paris: Economica/OST.

Amable, Bruno, Rémi Barré and Robert Boyer (1997b) 'Diversity, coherence and transformations of innovation systems', in R. Barré, M. Gibbons, John Maddox, B. Martin and P. Papon (eds), *Science in Tomorrow's Europe*, Paris: Economica International, pp. 33–49.

Amable, Bruno and Robert Boyer (1993), 'L'Europe est-elle en retard d'un modèle technologique?', *Economie Internationale*, 56, 4ème trimestre, pp. 61–90.

Aoki, Masahiko (2001), *Towards a comparative institutional analysis*, Cambridge, MIT Press.

Aoki, Masahiko (2001), 'Pourquoi la diversité institutionnelle va-t-elle persister et les institutions continuer à évoluer', to appear in R. Boyer and P. Souyri (eds), *Japon et Europe: Le second défi américain*, Paris: La Découverte.

Baslé, Maurice, Jacques Mazier and Jean François Vidal (1993), *Quand les crises durent . . .*, Paris: Economica (2nd edition: 1999) (1984), *When Economic Crises endure*, M.E. Sharpe Publisher, (1999).

Beffa, Jean-Louis, Robert Boyer and Jean-Philippe Touffut (1999), 'Les relations salariales en France: Etat, entreprises, marchés financiers', *Notes de la Fondation Saint Simon*, 107, June.

Bowles, Samuel and Robert Boyer (1990), 'Labour market flexibility and decentralisation as barriers to high employment? Notes on employer collusion, centralised wage bargaining and aggregate employment', in Renato Brunetta and Carlos Dell'Aringa (eds), *Labour Relations and Economic Performance*, London: Macmillan, pp. 325–53.

Boyer, Robert (1995), 'Training and employment in the new production models', *STI-Review*, 15, Paris: OECD, pp. 105–31.

Boyer, Robert (1998a), 'An essay on the political and institutional deficits of the euro. The unanticipated fallout of the European monetary union', *Couverture Orange CEPREMAP*, no. 9813, August. Published in Crouch Colin (ed.) 2000, 'After the euro', Oxford, Oxford University Press, pp. 24–88.

Boyer, Robert (1998b), 'Comment favoriser la coopération dans des sociétés conflictuelles? Quelques réflexions sur la transformation des relations professionnelles en Europe', in Henri Nadel and Robert Lindley (eds), *Les relations sociales en Europe: Economie et institutions*, Paris: L'Harmattan, pp. 151–82.

Boyer, Robert (1999a), 'Une lecture régulationniste de la croissance et des crises', in P. Combemale and Jean-Paul Piriou (eds), *Nouveau Manuel Sciences Economiques et Sociales*, Paris: La Découverte, pp. 290–304.

Boyer, Robert (1999b), 'Deux enjeux pour le XXIe siècle: discipliner la finance et organiser l'internationalisation', *Techniques financières et développement*, 53–4, December 1998–March 1999, *Epargne sans frontière*, pp. 8–19.

Boyer, Robert (1999c), 'Is a finance led growth regime a viable alternative to Fordism?', Mimeograph CEPREMAP, *Economy and Society*, February 2000.

Boyer, Robert (sous la présidence de) (1999d), *Le gouvernement de la zone Euro*, Rapport du groupe de travail du Commissariat Général du Plan, Paris: La Documentation Française.

Boyer, Robert (2001), 'La diversité des institutions d'une Croissance tirée plan l'information on la commaissance, "dans Centre Saint Gobain pour l'Economie" Institutions et Croissance' Paris, Belin pp. 325–54.

Boyer, Robert, Elsie Charron, Ulrich Jürgens and Steve Tolliday (eds) (1998), *Between Imitation and Innovation*, Oxford: Oxford University Press.

Boyer, Robert and Mario Dehove (2001), 'Du "gouvernement économique" au gouvernment tout court. Vers un fédéralisme à l'Européenne', *Critique Internationale* no. 11, pp. 179–195.

Boyer, Robert and Michel Didier (1998), 'Innovation et Croissance', *Les Rapports du Conseil d'Analyse Economique*, 10, Paris: La Documentation Française, pp. 11–132.

Boyer, Robert and Jean-Pierre Durand (1997), *After Fordism*, London: Macmillan Business.

Boyer, Robert and Yves Saillard (eds) (2001), *Régulation Theory: The State of the Art*, London: Routledge.

Caroli, Eve (1995), 'Croissance et formation: le rôle de la politique éducative', *Economie et Prévision*, 116, 1995–5.

Cohen, Daniel and Michèle Debonneuil (2000), 'La nouvelle économie', *Conseil d'Analyse Economique*, 28, Paris: La Documentation Française, pp. 9–49.

European Commission (1997a), 'Second report on science and technology indicators, 1997,' *Studies 1*, December, Luxembourg: printing office of European Community.

European Commission (1997b), *European Employment*, September, Luxembourg: printing office of European Community.

Conseil d'Analyse Économique (1998), *Coordination européenne des politiques économiques*, 5, Paris: La Documentation Française.

Davanne, Olivier (1998), 'Instabilité du système financier international', *Conseil d'Analyse Economique*, 14, Paris: La Documentation Française.

Dumez, Hervé and Jeunemaître, Alain (1996), 'The convergence of competition policies in Europe: Internal dynamic and external imposition', in Suzanne Berger and Ronald Dore, *National Diversity and Global Capitalism*, Ithaca: Cornell University Press, pp. 216–38.

Economic Policy Committee (1999), 'Synthesis report on structural reforms in Member States', Report addressed to the Council and the Commission, Brussels, 25 February, EPC/II/168/99.

Economie Européenne (2000), *Les grandes orientations des politiques économiques de 2000*, 70, Luxembourg: Communautés Européennes.

Esping-Andersen, Gøsta (1999), 'Ageing societies, knowledge based economies, and the sustainability of European welfare states', Mimeograph, Universita di Trento and Universitat Pompeu Fabra, September.

European Commission (1998), 'Wages and employment', EC/DGV – OECD/DEELSA seminar, Luxembourg: Office for Official Publications of the European Communities.

European Commission (1999a), 'Commission's recommendation for the Broad Guidelines of the Economic Policies of the Member States and the Community', Mimeograph Brussels, 30.03.

European Commission (1999b), 'Community Policies in Support of Employment', Mimeograph Brussels.

European Commission (1999c), 'Joint employment report 1999, Part I: The European Union', Mimeograph Brussels.

European Commission (1999d), 'Multilateral review of economic reforms in Member States, opinion addressed to the Council', Mimeograph Brussels, 5 August, EPC/II/444/99.

European Commission (1999e), 'Strategies for jobs in the information society', Report to the European Council (Draft of 17 September), Mimeograph Brussels.

European Community (1997), 'Survey on innovation', Mimeograph, Brussels.

European Council (1999a), 'Presidency conclusions of the Cologne European Council', 3–4 June.

European Council (1999b), 'Council resolution on the 1999 Employment Guidelines', Mimeograph Brussels, 29 July.

Freeman, Richard B. (1998), *Wages, employment and unemployment: an overview*, European Commission, EC/DGV – OECD/DEELSA seminar, Luxembourg: Office for Official Publications of the European Communities, pp. 21–31.

Freyssinet, Jacques (1999), 'Système de régulation et stratégies des acteurs', *Les Cahiers de l'Observatoire de l'ANPE: Les transformations du marché du travail*, 30 and 31 March, pp. 29–34.

Fujimoto, Takahiro (2001), 'La diversité de l'organisation des firmes: une analyse,' in R. Boyer, P. Souyri (eds) *Mondialisation et Régulations: Europe et Japon face à la singularité américaine*, Paris: La Découverte, pp. 131–7.

Giddens, Anthony (1998), *The Third Way. The renewal of Social Democracy*, London: Polity Press.

Gros, Daniel *et al.* (2000), 'Quo Vadis Euro: The cost of muddling through', Deuxième rapport du CEPS Macroeconomic Policy Group, Brussels.

Guterres, António (Presented by) (1999), 'A European employment pact for a new European way', Report adopted by the Party of European Socialists, Milan, 1–2 March.

Herzog, Philippe (1999), *Manifeste pour une démocratie européenne,* Paris: Les Editions de l'Atelier/Les éditions ouvrières.

Institut de l'Entreprise (1998), 'Croissance et emploi: pourquoi les Pays-Bas font-ils mieux que la France?', Séminaire organisé par l'Institut de l'Entreprise à l'initiative de la Commission 'Benchmarking' présidée par Bernard Esambert, Paris.

Jacquet, Pierre and Jean Pisani-Ferry (2000), 'La coordination des politiques économiques dans la zone euro: bilan et propositions', *Conseil d'Analyse Economique*, 27, Paris: La Documentation Française, pp. 11–40.

Kim, Haknoh (1999), 'Constructing European collective bargaining', *Economic and Industrial Democracy*, 20 (3 August), pp. 393–426.

Krugman, Paul (1990), *Economic Geography*, Cambridge, MA: MIT Press.

Majnoni d'Intignano, Béatrice (1999), *Egalité entre femmes et hommes: aspects économiques, Conseil d'Analyse Économique*, 15, Paris: La Documentation Française.

Marks, Gary, Fritz W. Scharff, Philippe C. Schmitter and Wolfgang Streeck (1996), *Governance in the European Union*, London: Sage Publications.

Maurel, Françoise (sous la présidence de) (1999), 'Marché unique, monnaie unique: trois scénarios pour une nouvelle géographie unique de l'Europe', rapport du groupe 'Géographie économique' Commissariat général du Plan.

Maurice, Joël (sous la présidence de) (1999), 'Emploi, négociation collective, protection sociale: vers quelle Europe sociale?', Rapport du groupe 'Europe Sociale', Commissariat général du Plan, Paris: La Documentation Française.

OECD (1999), 'L'économie fondée sur le savoir: des faits et des chiffres', Réunion du Comité de la politique scientifique et technologique (GPST) au niveau Ministériel, OECD, Paris 22–3 June.

Petit, Pascal (1998), 'Formes structurelles et régimes de croissance de l'après fordisme', *L'Année de la Régulation 1998*, vol. 2, Paris: La Découverte, pp. 169–98.

Quermonne, Jean-Louis (1999), *L'Union européenne en quête d'institutions légitimes et efficaces*, Rapport Commissariat Général du Plan, October.

Seibel, Claude (1999), 'L'évolution du marché du travail en Europe', *Les Cahiers de l'Observatoire de l'ANPE: Les transformations du marché du travail,* 30 and 31 March, pp. 19–28.

Soete, Luc (1999), 'The challenges and the potential of the knowledge based economy in a globalised world', Mimeograph MERIT, September.

Taddei, Dominique (1999), *Pour des retraites choisies et progressives*, Rapport pour le *Conseil d'Analyse Economique*, 21 Paris: La Documentation Française.

UNICE (1999), *Libérer le potentiel d'emploi de l'Europe. La politique sociale européenne à l'horizon 2000: les vues des entreprises*, Brussels.

Visser, Jelle and Anton Hemerijck (1997), *A Dutch Miracle,* Amsterdam: Amsterdam University Press.

World Bank (1999), *Entering the 21st century, World Development Report 1999/2000*, Washington DC.

6. International benchmarking as a policy learning tool

Bengt-Åke Lundvall and Mark Tomlinson

1. INTRODUCTION

The purpose of this chapter is to help build the foundation of processes of policy learning related to the Portuguese Presidency initiative on 'Economic growth and social cohesion – a Europe based on knowledge and innovation'. More specifically the aim is to give a critical assessment of the standard uses of benchmarking and to suggest a new kind of benchmarking process that overcomes some of the traditional weaknesses as a method of international policy learning. New policy 'architectures' are required for the European Union in the face of rapid developments in new technologies and the increasing complexity that these bring. In a world of rapid change, learning processes have to become more flexible and varied in order to cope with new features of socioeconomic life. We argue that naïve and mechanical applications of benchmarking procedures are highly problematic, because they undermine democracy and give rise to biased processes of institutional reform. But we also argue that the basic idea behind benchmarking – to stimulate processes of 'learning by comparing' – is sound and can be seen as a useful policy learning tool in the new environment.

We start out by analysing the origin of the concept 'benchmarking' in the private sector, but the main focus is on benchmarking involving international comparisons of institutions, economic indicators and policies related to competence building and innovation. We will refer to examples of what we will call 'naïve' benchmarking which are built upon simplistic ideas about the real world. There are also examples of the abuse of benchmarking, aiming at advancing narrow objectives of international business or technocrats in the public sector. However, we will argue that benchmarking, as a broadly defined process of comparing, evaluating and peer reviewing, does have a role to play in supporting policy learning. In a rapidly changing and complex world more 'precise' logical procedures – such as the use of full-blown econometric models – may be too rigid. In 'the learning economy' learning by comparing across national systems can be used to enhance the quality of policy making.

The reader should be warned from the outset that the proposed approach may appear to be more cumbersome and complex than methods used so far. We will try to take into account the major weaknesses with the present practises and especially we will emphasise the importance of taking into account the systemic character of national policies and institutions. In a sense, this chapter aims at 'benchmarking different benchmarking practices'.

Studying the literature on benchmarking, given the wide popularity of the concept, it is striking how little critical reflection there has been on the methodological problems involved. Even international organisations such as the OECD dominated by economists, normally well aware of parallel problems inherent in comparing productivity and welfare implications of economic policy, seem to become much more relaxed from critical reflection when they enter the field of 'benchmarking'.

In recent years the EU has been particularly keen on using benchmarking as a tool for European industry (see European Commission, 1999). The Commission has now also adopted a benchmarking ideology for the public sector:

> In an industrial and societal perspective, benchmarking is an essential policy instrument for improving the quality and the effectiveness of public services. Benchmarking should therefore not only be applied to private and public companies but also to hospitals, schools, universities, administrations, etc. (European Commission 1999:3)

The documents from the Lisbon European Council refer to the concept of benchmarking in several different policy areas. The purpose of emphasising these procedures is more far reaching than before. The Portuguese Presidency introduced to the European Council an institutional innovation referred to as 'the Open Method of Coordination' and one way this idea was made operational by national administrations and the different parts of the Commission was through references to procedures of international benchmarking. As can be seen from the introductory and concluding chapters the general aim of the summit was to advance the institutional adaptation of Europe to the new context of the knowledge economy and to stimulate European integration and convergence while respecting the principles of subsidiarity.

This points to a kind of benchmarking that is more ambitious, more overtly political and more complex. The idea is that benchmarking should contribute both to a convergence at the European level of national objectives and instruments and simultaneously promote innovation and social cohesion. When assessing the use of benchmarking in the European context, this quite complex agenda has to be taken into account.

The implementation of benchmarking is expected to take place differently in different policy areas. At regular meetings representatives of national governments will meet to consider progress made in specific areas of policy and it is assumed that this will also give inspiration for learning from each other's experiences. It is assumed that 'benchmarking' your own economy and using good practices from other countries as benchmarks is a way to stimulate progress toward shared objectives and instruments. In order to consider the potential usefulness and design of such a procedure it is useful to start with a basic analysis of how the concept 'benchmarking' was first developed in the private sector and later on adapted for use by the public sector.

2. WHAT IS BENCHMARKING?

Benchmarking has been defined in different ways in the management literature. Also there are different kinds of benchmarking mentioned in the literature and different ways of doing it. It always involves some sort of systematic comparison of one institution's outcomes or processes with either some other institution(s) or some accepted standard. It is usually a rigorous comparison with an institution that is regarded as superior in performance. Key actions within all benchmarking definitions are search and improvement.

Bogan and English (1994) are responsible for one of the early definitions which gets cited frequently in management literature:

> Benchmarking is the systematic observation of organisational routines and the comparison of performance with superior units at the levels of resource use and efficiency and effectiveness (inputs and outputs).
> Benchmarking is the search for industry best practices that lead to superior performance.

It is thus basically about comparing one company's performance with another who is considered one of the best, if not the best, in the field. A learning process takes place whereby the benchmarking company can adjust its behaviour based on observation of the benchmarked company and thus improve its efficiency. Crucially, data are generated that can be compared between organisations. Several other definitions have also been put forward (see, for instance, Table 6.1).

Table 6.1   Different definitions of benchmarking cited in Cox, Mann and Samson, 1997: 287

---

(1) A continuous, systematic process for evaluating the products, services and work processes of organisations that are recognised as representing best practices for the purpose of organisational improvement (Spendolini, 1992)
(2) The continuous process of measuring products, services and practices against the toughest competitors or the companies recognised as industry leaders (D.T. Kearns, CEO, Xerox Corporation, cited in Camp, 1989)
(3) A continuous search for, and application of, significantly better practices that lead to superior competitive performance (Watson, 1993)
(4) A disciplined process that begins with a thorough search to identify best-practice organisations, continues with the careful study of own practices and performance, progresses through systematic site visits and interviews, and concludes with an analysis of results, development of recommendations and implementation (Garvin, 1993)

One of the first documented examples of benchmarking which had a significant effect on this literature is the famous Xerox example from 1980. In this case a systematic comparison of photocopier manufacturing by Xerox in the US with Fuji-Xerox in Japan led to a more detailed and formalised comparison across the industry. This was in tandem with the actual codification of Xerox's production routines for the first time (see Bogan and English, 1994; Boxwell, 1994; Macneil *et al.*, 1994)

3. BUSINESS BENCHMARKING IN THE LEARNING ECONOMY

Benchmarking then as it has been developed in the private sector has been used as a method to enhance the performance of private firms. The basic idea is that some firms perform better than others and that the performance reflects firm-specific 'practices' that can be transferred from one firm to another. A systematic mapping of which firms are doing best among a population of comparable firms and what practices those firms use (best practice) generates ideas about how to become better. By comparing yourself with the best, you get an idea about how to enhance performance.

This basic idea is less trivial than it appears at first sight. For instance, it stands in sharp contrast to standard microeconomics where it is assumed that all firms have equal access to technologies and modes of organisation. Neither is it easily reconciled with macroeconomic models assuming rational expectations. The popularity of the model may actually be seen as a confirmation of the fact that practitioners do not believe in the basic assumptions of standard models of economics.

It may also be seen as a reaction to increasingly turbulent competition where change is accelerating while at the same time new technologies become more complex and markets more volatile. In this 'learning economy' it is the capability to learn that determines the outcome of competition.[1] This is why management is eager to find ways to speed up the transformation toward more efficient methods. More sophisticated, logical and comprehensive methods, such as operational research, linear programming and technology assessment, may be too cumbersome and too slow when it comes to cope with a much more fluid, rapidly changing and uncertain reality. In such a context a more intuitive and interactive procedure may be more efficient. Instead of calibrating explicit and exact models, benchmarking will reflect the establishment of consensus on an incomplete, implicit and intuitive model.

In many situations this is probably a more useful approach than the ones depending on complete models and complete information. There is, of course, some risk in this movement toward a 'post-modern' management technique.[2] Benchmarking may, since it is often based on comparing numbers, appear to be a more exact technique than it is. It may be manipulated by those in power to impose controversial changes by referring to its analytical character. On the other hand the process of designing and implementing benchmarking may be a way of handling conflicts of interest and to reach consensus. Here the distinction between technocratic top–down and democratic bottom–up approaches is critical.

4.   DIVERSITY AND BENCHMARKING

There are many problems with simplistic versions of benchmarking when applied in the private sector. The idea that there is one single best practice that can be referred to as the benchmark is valid only under some very specific conditions.[3] To take a well-known example from the history of technology, the transfer from sailing techniques to steam power was not as simple as just moving from an old inferior to a new and much better practice. The two technologies co-existed for many years because the pressure from the new technology stimulated substantial progress in the old. Even today, for some specific purposes sailing might become the better technique. To take a more modern example, it is far from clear at present whether the supercomputer is obsolete. There is debate in computer science as to whether the combined power of large networks of PCs are actually better than high-powered mainframes. Fashions keep changing and there are often swings from one computing paradigm to another and back again. What is true for technologies is probably even more the case when we focus on organisational procedures, new marketing technique and competing management information systems. Here the human and social context and not least the educational background of employees may determine what is good and bad practice.

A second problem has to do with the very idea of copying. It is 'the lemming effect'. Stardom among management styles and firms is changing quite rapidly and some of the firms characterised as parading 'excellence' one year may be treated with disdain the next. Copying the stars may bring a whole population of firms on to the wrong path. Similar mechanisms are at work when it comes to international benchmarking of policies and institutions. In the beginning of year 2000, when the Lisbon European Council took place, the US economy led in terms of being the most successful 'new economy' and this was much more obvious than it was at the end of the same year. The deroute of the Nasdaq market has demonstrated the instability inherent in a system built on exaggerated technological expectations and debt-financed speculation.

Regardless of whether it is possible to define what is best practice, too much copying may always be problematic. Even if there were a rather clear best-practice firm to copy from, the copying process itself may undermine the dynamic capabilities of the industry as such. One of the most important new insights coming from the research within evolutionary economics is that 'diversity' is a key to economic evolution, innovation and economic growth. Benchmarking might be seen as intensifying the selection process that market competition gives rise to and it might result in too little diversity and heterogeneity. In a national industry, where all firms in an industry use the same routines and procedures, firms have less to learn from each other than one where there is more diversity in these respects.

This kind of effect can perhaps be most clearly seen in financial markets where the diffusion of similar procedures create greater instability, but it has relevance also for many other areas; a classical illustrative example being the one about the fishing industry in Canada (Allen, 1988) showing how the co-existence of different categories of agents with different rationalities is a pre-requisite for a sustainable fishing system.

A similar point is often made in the context of the European integration process. In the European discourse 'diversity' across member states should be regarded as a potential source of innovation and growth, and 'valorising diversity' has been pointed to in contrast to strategies aiming at convergence of institutions and policies (see Boyer in this volume). Still it is true that on balance the single market project and the strong focus on reducing barriers to competition has stimulated processes of convergence. To a certain degree this reflects the strong emphasis on promoting European-wide competition policy. The distinction between legitimate national differences and barriers to the free movement of goods, services, labour and capital is becoming less clear as competition policy is extended to cover services and procurement procedures related to the public sector. The design of benchmarking procedures that follow on from the Lisbon European Council need to be explicit in recognising this European dilemma.

5. BENCHMARKING IN SYSTEMIC CONTEXTS

The last critical point to bring up is the most important and perhaps the most obvious. The context – defined in its economic, technical, geographic, historical and cultural dimensions – has a great influence when it comes to determine what is a best-practice way of doing things for a specific firm. What kind of raw materials do you have access to? What is the skill level and the culture of the engineers that work in the firm? How strong is the trade union and what is the average mobility of labour in the local labour market? What characterises a firm's network position in relation to suppliers, customers and knowledge institutions? What is the dominant mode of government regulation in relation to environmental issues?

Most of these factors are localised and cannot easily be changed. The literature on national competitive advantage (Porter, 1990), national business systems (Whitley, 1996) and national systems of innovation (Lundvall, 1992; Nelson (ed.), 1993) has as a common message that national differences in these respects are systemic. There is a similar analysis of regional systems (Maskell *et al.*, 1998). To neglect the local, regional and national context when selecting and introducing new ways of doing things would be highly problematic, and to benchmark only those firms that have all these conditions in common would, alternatively, reduce the number of comparable units to a handful or even to zero.

Given these critical points, one might ask, why does management in advanced firms bother about benchmarking at all? It seems to lead to more problems than it solves. There are at least two reasons why benchmarking may be useful as it is practised in real life. The first reason is that any attempt to get the members of an organisation to focus on performance and on what factors may help to make it better is useful in a world where things change very rapidly and where old routines become obsolete. Benchmarking obviously serves this general purpose. The basic idea that comparing your own way of doing things with that of others is one way of stimulating reflection. And moments of reflection are key elements in learning processes. Reflection enhances cognitive insights as well as competencies to be used in practice.

Secondly, most experienced and competent managers would avoid promoting the kind of naïve benchmarking that was sketched above. They would not, for instance, uncritically copy what seems to be best practice at a given moment. On the contrary, they would normally take into account the specific context they are operating in and adapt the so-called best practice according to circumstances and take into account how it affects flexibility in relation to future shocks and challenges.

6.   BENCHMARKING AS A MEANS TO PROMOTE THE FORMATION OF LEARNING ORGANISATIONS

One area where benchmarking, if designed intelligently, has great potential is related to the movement towards functionally flexible organisations and towards learning organisations. It is interesting to note that there are strong indications that certain generic characteristics of organisations seem to promote performance in a broad range of sectors, in this period characterised by rapid change and strong learning requirements.

According to recent research based on new data constellations, there is a tendency for 'functionally flexible' organisations to perform better than the average in different respects. They are more productive, they create more jobs and they introduce more product innovations. But since they are more exposed to competition than the 'functionally rigid organisations' they may not always be more profitable (Coriat, 2001).

But the same research also shows that the performance impact of introducing the new forms of organisations differs between sectors. In sectors such as construction and transport the context is such that organisational change does not promote performance to the same degree as it does within manufacturing and business services (Lundvall and Nielsen, 1999). Also the need for organisational change is stronger in big than in small units of production. So, in this case too, there is a need to be careful and to avoid defining generic best practices once and for all.

7.   COMPETITIVE VERSUS COLLABORATIVE BENCHMARKING

Cox, Mann and Samson (1997) point out that there may be certain power relations involved that are not addressed much in the literature. They use the concepts of collaboration and competition to reveal that there are often differences in the orientation of the benchmarking organisation and the benchmark organisation. Whereas collaborative benchmarking is about learning and involves a joint effort, competitive benchmarking is usually about superiority and rivalry. The former is driven by (ideally mutual) comparisons between organisations while the latter is usually an external standard-driven approach. However, there are legal and ethical issues involved in any type of collaboration and this applies to benchmarking as well (Spendolini, 1992; Liebfried and McNair, 1992).[4]

Thus many organisations do not want data about themselves made available to others. This has in part led to the development of private and public organisations that provide a service whereby they keep a database of firms' performance indicators and charge a fee to inquisitive firms for a comparison to be made with these data. The Department of Trade and Industry (DTI) in the UK has such a service for small and medium-sized firms. The 'UK Benchmarking Index' allows a company to measure its performance against others with respect to over sixty variables relating to finance, operations, management and business excellence. Industrial sector, size of company and geographical region can also be taken into consideration (see DTI, 1998).

In this context it is interesting to note that the kind of benchmarking aimed at in the wake of the Lisbon European Council is collaborative. The basic idea is that all countries will present comparable data and information about performance and policy initiatives. As we shall see later, even with a maximum of good intentions this might be difficult, since what appears to be the same variable may have a different meaning and content in different national contexts. For instance the common terminology of primary, secondary and tertiary education may cover different forms and contents in different countries.

8. SUMMING UP SOME OF THE PROBLEMS WITH BENCHMARKING IN THE PRIVATE SECTOR

Thus there are several fundamental problems inherent in the comparison of the routines and performance of firms/organisations. Benchmarking implicitly relies on some sort of objective measure of performance (normally referred to as 'best practice'). Taken out of context the observed routines or the outcomes of routines may not be strictly comparable. An institution doing 'badly' on one indicator may be doing rather better on another. The selection of indicators is thus a highly 'political' process and there will always be a tendency to give more weight to what is most easily quantifiable and measurable. Furthermore, indicators used to benchmark may be calculated using different assumptions. What is good practice will depend on the context. This begs the question just how sophisticated the techniques need to be in order to be useful?

9.   PUBLIC SECTOR BENCHMARKING

In recent years, generally at the request of national governments, public sector organisations have also started to use benchmarking. The European Commission has offered a cautionary note with respect to using a private sector methodology in the public sector (e.g., with respect to health systems):

> On the effectiveness of health systems in general, account must be taken of the difficulties of using techniques initially developed for the private sector. The competitive pressures are not the same. The objectives of public services in general are not defined by competition and consumers but through the demo-cratic process. (EC, 1999: 16)

Most of the problems we have referred to above reappear when we focus on benchmarking in the public sector and new ones are raised. While the idea that firms have complex objectives reflecting a compromise between different interest groups (owners, management, employees, customers, society at large, etc.) may be controversial, it is obvious that public activities normally will have to take into account conflicting interests and objectives. If it were not so there would be little reason to involve the public sector in organising the activity. When it comes to public activities motivated by falling marginal costs and externalities, economic efficiency may be the key issue. This may be true for constructing bridges, roads and telephone systems, but even here the issue of taking into account 'public access' will restrict the choice of 'best practice'. To locate such activities in the market would normally increase inequality both in terms of living standards and in terms of opportunities.

This is even more the case when the activity involves a strong element of human interaction. The quality of education, health and social services cannot be assessed exclusively by comparing input and output variables. In these areas benchmarking needs to be sensitive to the wider impact of procedures on human interaction and on the subjective satisfaction of service providers and service receivers.

There are of course many activities in the public sector that might be made more efficient without undermining social cohesion but there is a need to have a sensitiveness to the social and human dimension when pursuing benchmarking in those areas. It seems as if most benchmarking exercises pursued so far in the public sector have been initiated by technocratically oriented institutions and with a focus on financial variables. It is characteristic that the most ambitious benchmarking exercise so far at the European level has been the close surveillance of certain financial variables in connection with EMU cooperation. We see the recommendations of the Lisbon European Council as pointing to a new perspective where much stronger weight is given both to economic dynamics and to the social dimension.

The distinction between a *top–down approach* (where benchmarking is applied externally by some outside body such as a government ministry which can set performance targets) and a *bottom–up approach* (where individual organisations do the benchmarking themselves and find other organisations to compare themselves with) is especially important when analysing benchmarking in the public sector. Public institutions need to reconcile the interests of political and administrative leadership with the interests of employees, citizens and customers. Benchmarking imposed unilaterally by the central administrative leadership – such as ministries of finance and public administration – will not necessarily mobilise local forces positively in processes of change. On the other hand, there is nothing that excludes the idea that benchmarking could become an integral part of a democratic process and contribute to the formation of a European identity (Castells in this volume).

An interesting alternative to benchmarking is the establishment of *development contracts*. Here a central authority gives decentral units the opportunity to develop their own development plans, including objectives and instruments as well as a choice of benchmark indicators and benchmark standards. This is followed up by a surveillance of the implementation of the plans. A similar procedure could be established in connection with the Lisbon European Council follow-up. It could have built into it systematic comparisons of performance indicators and of policy strategies used by member countries, but it would explicitly take national priorities as its starting points. Such a procedure may have a number of advantages in terms of legitimacy and commitment as compared to more top–down approaches.

In the context of international benchmarking related to innovation, knowledge and social cohesion these distinctions are important. It is important to agree on what kind of *results* should be considered as the critical performance indicators. These need to balance the strengthening of the knowledge base and competitiveness, on the one hand, and distributional issues, on the other hand. Indicators reflecting income and knowledge creation need to be presented as distributions around a mean rather than just as an average value. Ideally there should be indicators of social distribution, gender distribution and regional distribution of the different indicators.

It is not obvious that each country will give the same priority to different performance indicators. Some of the richest countries might give stronger weight to environmental objectives while some of the small Nordic welfare states may be especially committed to an equal income distribution. Big European countries may have stronger ambitions in terms a broad and strong knowledge base while the small ones need to be much more specialised. Instead of hiding such differences behind vague statements of objectives that all can accept, it might be a good idea to get each national government to make their objectives explicit in 'development plans'.

If there are legitimate problems in connection with agreeing on common goals the identification of 'best-practice' *processes* is even more difficult. This is an area that is strongly politicised. The classical debate between those favouring state intervention and those believing more strongly in what markets can do is still alive and it has developed differently in different European countries (see also Table 6.2 for examples of how benchmarking may be used to promote specific political ideas). Actually, we believe that the focus on knowledge and innovation initiated by the Portuguese Presidency might be helpful to open up less-prejudiced international and national debates on these issues and to overcome some of the old differences between a British marketeer approach and a Continental regulationist approach. But again, given these historically rooted differences, it might be useful to give national governments and institutions a chance to formulate their own intentions also in terms of instruments rather than to impose on member countries what is supposed to be 'best practice' in policy-laden areas.

The kind of benchmarking indicated by the Lisbon European Council is one where the emphasis is on continuous processes of *improvement* rather than just on *evaluation*. But of course some kind of evaluation is necessary in order to promote improvement. The idea that the different national governments act as each other's critical evaluators is important for several reasons. The evaluated government needs to argue its case among equals and therefore also to reflect upon why they prefer to link certain instruments to certain objectives. The detailed evaluation of the specific country will force evaluators to try to understand the logic (and the incomplete models) behind national development plans. In both cases it is possible to foresee policy learning that increases the competence base of policy making and to get closer to a process of benchmarking where some true best practices may be identified and diffused among the European countries.

Table 6.2   The rankings of national systems by private organisations and their political impact

---

OECD, IMF, the World Bank and several Branches of the UN present annual statistics that make it possible to compare countries in terms of macroeconomic performance and structural characteristics. Policy makers will follow this flow of comparative data and inspire new policy initiatives.

A different type of benchmarking focuses on framework conditions for private enterprise and here a number of different indicators are weighed together into a single index and countries are ranked accordingly. Here it is private or semi-private organisations that pursue the exercise, and in the selection of indicators and weights there is ample room for ideological bias. Some of the best-known examples of such business-oriented rankings are:

1. *The World Competitiveness Yearbook* is published by the Swiss Management school IMD. It is constructed from 220 indicators grouped into eight categories.
2. *Global Competitiveness Report* is published by the World Economic Forum and is more based on survey data than the *Yearbook*. The overall ranking is based upon 170 indicators grouped into eight categories.
3. *Economic Freedom of the World* is published by the independent think-tank The Fraser Institute that has as its aim to promote the use of 'free markets' in the world. It brings together twenty-five indicators grouped into seven broad categories.

Characteristic of all three indexes is that a big public sector is regarded as something that weakens the competitiveness of a national economy and that a high degree of openness, in terms of international trade, contributes to a high ranking. The ranking procedure may be seen, at least partially, as a political intervention aimed at affecting national policies in such a way that the interests of international business are taken into account. This will be especially successful when there is a spill-over from these exercises to the benchmarking pursued by public sector institutions.

---

10.   BENCHMARKING NATIONAL SYSTEMS OF INNOVATION

The concept 'innovation system' was introduced in the middle of the 1980s (Lundvall, 1985: 55) to capture the relationships and interactions between R&D laboratories and technological institutes, on the one hand, and the production system, on the other hand. The first widely diffused publication that used the concept of a 'national system of innovation' was the analysis of Japan by Christopher Freeman (1987). The concept was definitely established in the innovation literature as a result of the collaboration between Freeman (1988), Nelson (1988) and Lundvall (1988) in the collective work on technology and economic theory (Dosi *et al.* (eds), 1988).[5]

The most fundamental reason for innovation scholars to begin to think in terms of innovation systems had to do with the fact that it was increasingly realised that innovation is an interactive process. Through the 1970s and the 1980s the linear model, where new technology is assumed to develop directly on the basis of scientific efforts, and, thereafter, to be materialised in new marketed products, was discredited by empirical work. The Sappho-study pursued by Freeman and his colleagues at SPRU in the beginning of the 1970s (Rothwell, 1977) had already given strong support to the idea that success in innovation has to do with long-term relationships and close interaction with external agents. The presentation of 'the chain-linked model', by Kline and Rosenberg (1986), was important because it gave a specific form to an alternative to the linear model.

All this constituted the first step toward the development of the idea of national innovation systems. The second step was to realise that the relationships and interactions between agents had to involve non-market relationships. These relationships were presented as 'organised markets' with elements of power, trust and loyalty (Lundvall, 1985) and presented as the only possible solution to the conundrum of product innovations.[6]

The third step was to realise that national contexts offered quite different possibilities for establishing organised markets. A series of studies pointed to the long-term and selective interfirm relationships in Japan and contrasted them with those of the Anglo Saxon countries which were more of the arm's-length type (Dore, 1986; Sako, 1990). Other important national differences, such as the form and content of the interaction between universities and industry, the education and training system and the kind of interaction it fostered among specialists, the financial markets, etc., that had been described separately in the literature, were gradually taken into account and entered into the system's perspective.

A system's approach to innovation makes it more complex to interpret the results of benchmarking and to define action on the basis of benchmarking. For example, it does not support strategies that have as their exclusive aim to increase the R&D budget. Different types of competence matter more or less, dependent on industrial structure and institutional set-ups of the national economy. Other factors have to be taken into account, such as the demand side and the user competence. Crucially, what matters most are relationships between the different elements of the system and it is often extremely difficult to get good indicators for 'connectivity' and social capital. Having excellent universities does not help if the interactions with competent users of academic research are weak. In spite of these complications the innovation system approach has diffused rapidly to national and international policy-making bodies, such as OECD, the European Commission, UNCTAD and the World Bank. Many bodies have integrated the concept into their policy analysis, while a number of national studies have inspired new, more integrated policy strategies aimed at improving innovation and competitiveness.

This has led to a kind of benchmarking done at the NSI level. Organisations such as the OECD publish supposedly comparable indicators for several countries and with respect to several parts of the system (such as labour market indicators, R&D spending, GNP, training, health, exports or whatever; see for example OECD, 1999). Lengthy discussions and procedures are developed to ensure comparability between countries and there is a good deal of cooperation between nations as to what the best indicators are. From time to time detailed national surveys with international examiners are pursued and results are presented to government experts from other countries. These measures are in one way or another used by policy makers or economists to assess relative performance and propose tentative solutions where indicators show poor functionality. Pursuing benchmarking while taking into account such a systemic perspective reduces the risk for misinterpretations of assessing single indicators and for disregarding specificities in the national context.

Academic studies such as the one by Amable, Barré and Boyer (1997) have gone quite a long way in characterising different national systems without referring explicitly to benchmarking. The types of innovation projects funded under the EU TSER programme mostly fall into this category of pragmatic mapping and comparing systems and elements of systems. There is a growing body of cross-national research being undertaken under the umbrella of a national systems perspective, but there is a lack of systematic coordination when the research is used to guide the creation of effective policies.

Another problem with international quantitative comparisons is that indicators are quite weak in some of the most important aspects of innovation systems. This is true for measurements of intensity and quality of interaction and network formation and for the presence of functionally integrated learning organisations in the private economy. It is also true for the formation of tacit, experience-based, knowledge – a factor that becomes increasingly important in the learning economy (Polanyi, 1958/1978; Polanyi, 1966 and OECD, 2000). Finally, it is difficult even to conceptualise certain elements that have become crucial for sustainable economic growth, such as 'trust' and 'social capital'. In all these areas there is a need for fundamental efforts to establish more conceptual clarity and to advance toward indicators that can be compared across national borders.[7]

11.   ON THE NEED TO FOCUS ON THE NATIONAL SYSTEM FOR KNOWLEDGE CREATION AND LEARNING

These two last points indicate a need to broaden the innovation system concept. We see two major challenges for the future work on national systems in the context of economic growth and development. The first has to do with the need to extend the analysis so that it covers all aspects of competence building including formal education and competence building in learning organisations. The second has to do with the need to broaden the analysis of economic development and to see how the different forms of competence building are conditioned by social considerations. These challenges coincide with the new policy directions initiated by the Portuguese Presidency at the Lisbon European Council by evoking the theme of a Europe based on knowledge, innovation and social cohesion.

The most important elements in current innovation systems have to do with the learning capability of individuals, organisations and regions. The very rapid rate of change gives a premium to those who are rapid learners. This is reflected in the forms of organisation inside firms, new mixtures between cooperation and competition as well as in new forms of governance. It presents all organisations specialised in the production, diffusion and use of knowledge with new challenges.

So far the studies of national systems of innovation have given too little emphasis to the subsystem related to human resource development. This includes formal education and training, labour market dynamics and the organisation of knowledge creation and learning within firms and in networks. This subsystem will be confronted with very strong needs for social invention in the near future in all national systems and quite a lot of the peculiarities of national systems are rooted in this sub-system.

Another new focus must be on the part of business services that specialise in producing, gathering and selling knowledge. This sector is growing more rapidly than any other sector and new empirical studies indicate that it is becoming a key sector in the French structural school sense. More and more producers of tangible products and traditional services are moving into this field. To understand how such businesses operate within and across national borders is another key to understanding future economic dynamics.

The production and diffusion of knowledge is itself changing character. Some elements of knowledge become codified and much more mobile globally, while other key elements remain tacit and deeply embedded in individuals and organisations and localities. To understand these processes better is a key to establish a new kind of economy. It has been argued by

Marx that what really made the industrial revolution a revolution was not the use of machinery, but rather the stage where machinery was used to produce machinery. It may be the case that it is only when we can systematically apply knowledge to the production of knowledge that we will witness the establishment of the learning economy.

It is also important to note that 'social capital' (Coleman, 1988; Putnam, 1993; Woolcock, 1998) might become the most important scarce factor in the future. Historically this kind of resource has constituted the major comparative advantage of certain small nations (Kutznets, 1960) – and also of some regions within nations as in some of the Italian industrial districts. It is a pre-requisite for learning but it is constantly threatened in the globalising learning economy. Globalisation undermines traditional communities based on national belonging, religion, tradition, class, locality and family. If left to itself the globalising learning economy gives rise to increasing inequality between people and regions. To find ways of reproducing and creating new forms of social capital is a key task for governments and for the European Union. To integrate indicators of social equality and of social capital when benchmarking innovation systems is a major challenge.

An interesting example is the recent plan for industrial development in Denmark endorsed by nine different ministers. Here the first step in the benchmarking proposed is one where GNP/capita is benchmarked together with indicators on social cohesion and ecological sustainability. The second step is to follow a number of specific indicators in the three fields and to compare them with international data.

12. TOWARD A MORE SYSTEMATIC PROCESS OF BENCHMARKING INNOVATION SYSTEMS

It might be useful to consider how benchmarking innovation systems could be turned into a more systematic and rigorous procedure. If we can represent benchmarking as a sequence of procedures to be adopted, it might look like Figure 6.1. One basic idea is to make sure that there is a continuous dialogue both between national governments and the European community and between the national government and their own citizens. A major risk in the present era is a growing alienation between a minority of sophisticated cosmopolitan modernists and a majority that is sceptical to globalisation and to the further integration of Europe.

Basically, at some point there must be *a planning and auditing stage* to outline which particular areas of the NSI are to be examined and which have the most pressing problems. Government and academic agencies can perform internal research and assessment in order to fully grasp the nature

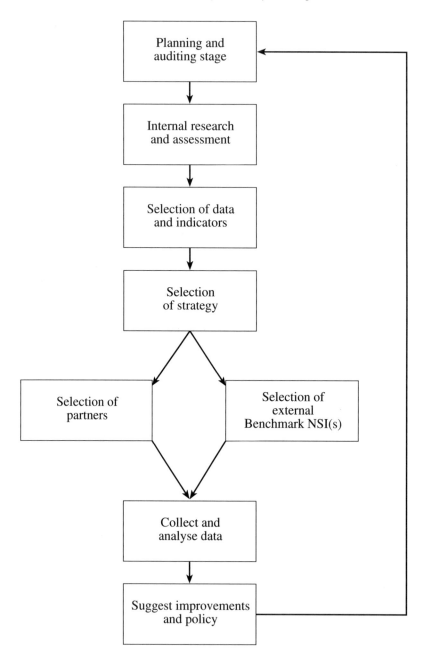

Figure 6.1 A schematic model of the NSI benchmarking process

of the problems with subsystems of the NSI and then select relevant indicators and data sources that might be useful. In doing so they may take international documents such as the ones coming out of the Lisbon European Council as their starting point. New communication techniques make it possible to organise already at this stage a number of small gatherings of experts, interest groups and ordinary citizens (panels) to discuss what are the most important challenges. A dialogue between laymen and experts may aim at reaching consensus regarding what performance and process indicators to focus upon.

At this stage the benchmark strategy will be adopted. The strategy should define the objectives of the exercise and the procedures to follow. Procedures should take into account the need to learn continuously both in the international and the national context. A European context would present national governments with a number of co-existing systems and some of these may be close enough in terms of objectives and instruments that a more strict benchmarking procedure may be established. But to begin with the confrontation and exchange around national development plans may be less ambitious until the international process makes it possible to achieve greater comparability.

In this case some external indicators may be used as a guide to set some form of realistic target. Of course, a combination of these strategies might be pursued. Once the strategy is set, *indicators and data can be analysed* within the framework proposed at the planning stage.

This should lead to policy recommendations and suggestions where improvements to the system can be made and crucially monitored. Then we return to the beginning again and re-assess, re-audit, see whether our policies have led to any improvements and begin the process over again. It would be important to establish permanent national panels of experts and non-experts to follow each step in the process and to give their advice on how to proceed. Democratic procedures have their own rationale and they also increase the probability that the outcome will be strategies with a good chance of being implemented.

Obviously this is an idealised version of what is possible and there are several caveats to be considered. What is of particular importance is the issue of whether NSIs are readily comparable at all. As has already been pointed out, in relation to some of the most important relationships and resources, current indicators are extremely crude. For benchmarking to be 'intelligent benchmarking', the indicators must be set within the national/regional/global context within which they are embedded. A careful analysis of many indicators will have to be undertaken. Issues and problems of comparability will be accentuated if a benchmarking methodology is adopted between nations. For instance, there must also be a consideration of what a relevant benchmark NSI actually is. Is there any point in comparing Ecuador with the USA?

Also even indicators that may appear relatively unproblematic on the surface will turn out to have hidden dangers in interpretation. For example, some studies in the UK suggest that as many as 20 per cent of graduates do not actually need their degree for their current job. See Green, McIntosh and Vignoles (1999) for a review of this literature. Whether we agree with this or not, it shows that an uncritical acceptance of indicators relating to education and training in the UK may give a biased view of the 'success' of the higher education system in Britain.

Similarly the move towards 'better' systems has to be treated carefully. There may not be one best method of organising an institution within an NSI due to cultural or historic factors, etc. and the interrelationships between different parts of the system have to be considered. The labour market may be the single area where we find most examples of ideologically biased and naïve benchmarking. The benchmarking of Danish labour market character-istics undertaken by the OECD and also by the Danish government is an interesting case.[8] It has repeatedly been pointed out that the substitution rate and duration of unemployment support is much too 'generous' in Denmark; and it is true that it stands out in these respects when compared to other coun-tries. At the same time most labour market performance indicators are out-standing for Denmark. Participation rates are extremely high, structural unemployment is low and so are both long-term and youth unemployment. And, not least, mobility is on a par with or even higher than in the US. The focus has been on the impact of 'generous' unemployment support rates for registered unemployment – it is neglected that in the Danish case the more important impact may have been the extremely high participation rate of women and the high and freely chosen mobility. This may be seen as one important aspect of the employment bias of the Nordic Welfare Regimes (Esping-Andersen in this volume). It also illustrates the problem with focus-ing on one single variable at a time and neglecting systemic features. Recent work on 'employment systems' (Marsden, 1999) gives a better basis for 'learn-ing by comparing' in this field and to use benchmarking in an intelligent way.

This example illustrates the danger in imposing general best-practice benchmarks without taking into account the whole socioeconomic system. Some pilot studies of benchmarking labour market indicators have raised some of these issues already. A study that tried to adapt a business bench-marking approach (using radar charts) to EU labour market performance noted that:

> The substitution of other indicators that are equally plausible in terms of the
> European employment strategy (e.g. promotion of self-employment or integra-

tion of handicapped persons) might lead to somewhat different comparative results. Moreover, the quantitative indicators actually used are inevitably only approximations because of the institutional and cultural diversity in the employment systems compared, and the qualitative dimensions of indicators (e.g. of employment) are neglected due to the lack of agreed measures. (Mosley and Mayer, 1998: 26)

In this study using some indicators Japan and the USA came out worse than the EU, while when using others the reverse was the case. The rankings of different countries were very sensitive to the indicators chosen. This highlights the need to know what the object of the exercise is. Indicators should be chosen with very specific goals and questions in mind.

Another study noted that benchmarking used by itself may be of limited value. A cross-national pilot study benchmarking skills undertaken for DGIII identified several shortages of certain kinds of skills (notable interpersonal and communication skills) within member states but still concluded that:

> the benchmarking technique as it is normally conceived is not appropriate to handle the skills issue in a broad way. It is too restraining since any quantitative indicators of skills performance would not be precise and would leave out much qualitative information which is very important in this field. . . .
>
> In order to assess the skills problem in European industry, instead of using the standard benchmarking tool, we suggested a more flexible comparative approach. Such an approach would be qualitative and multiform. (European Commission, 1998: 115–16)

They advocate the use of case studies and more qualitative techniques rather than an exclusive reliance on quantitative indicators. In other words, several different techniques and methodologies will need to be implemented in order to make international benchmarking a successful policy learning tool.

13.   INTELLIGENT BENCHMARKING

So where do we go from here in terms of advocating a superior form of benchmarking as a tool for policy recommendation? It seems that what has been said leads to the following primary considerations. Firstly benchmarking should not be seen primarily as a narrow technical procedure focusing on comparing quantitative data. Its rationale should be that it focuses attention on the efficiency of a system, stimulating reflection and thereby supporting learning among those involved.

The procedure of establishing and pursuing the benchmarking exercise should take into account learning taking place at different levels, that is at the European as well as at the national level. Elements of bottom–up processes where panels of citizens and national experts get a chance to follow the different steps from planning to policy implementation are necessary, in order to avoid polarisation and a technocratic bias. This implies that it is preferable to have the maximum of openness and cooperation between benchmarking countries. A mutual learning environment will be the most fruitful.

In the beginning it might be useful to develop less-demanding forms of comparative analyses than benchmarking in a strict sense. Gradually it might be possible to develop a common understanding and shared objectives that makes it more meaningful to benchmark specific aspects of the innovation system. Related to this is the fact that when delimiting the field to be covered by the benchmarking exercise, it is important to avoid a narrow focus on single variables. It is important to include in the benchmark exercise performance indicators referring to economic dynamics and indicators referring to distributional issues. Before drawing conclusions from benchmark exercises it is important to take into account the systemic features of the national economy.

Benchmarking innovation systems, learning effects and social cohesion involve the development of more reliable indicators for the quality and intensity of relationships, interactions and networks. The same is true for indicators referring to the characteristics of learning organisations and experience-based tacit knowledge. In these and other areas there is even a need for new conceptual work before meaningful indicators can be constructed. This is especially true for 'social capital' that is a crucial element in the formation of intellectual capital. It may be that a complete overhaul of the traditional indicators used for assessing competitiveness may have to take place. The business model of benchmarking may be less appropriate for parts of the system that involve intense human interaction and the formation of social capital than for systems where the interaction with citizens is communicated through techno-economic infrastructures. In the first category introducing bottom–up and democratic procedures is especially important in order to obtain learning effects and successful implementation.

The benchmarking approach is based on incomplete and intuitive models of reality and it may be especially adequate in the present era of complexity, rapid change and uncertainty. To use simple analytical tools such as operating with intuitive and very incomplete models of reality may be a more realistic strategy than aiming at more ambitious grand econometric models. But there might be alternatives in between these two extremes. Using simple statistical techniques to map systemic characteristics may actually be helpful in order to avoid naïve and ideologically biased benchmarking. Learning by comparing, includes the international comparison of indicators, the use of simple statistical techniques to map causalities and the qualitative comparison of systems. Used in such a context, benchmarking has the potential to significantly improve the effectiveness of institutions within national innovation systems by creating a mutual learning environment.

POST SCRIPTUM ON BENCHMARKING AND THE OUTCOME OF THE PORTUGUESE PRESIDENCY

At the Lisbon European Council benchmarking and the search for 'best practise' was referred to in connection with almost all major policy areas. It was referred to as an important element in the open method of coordination in relation to the development of the information society, enterprise policy, structural reform, education policy, research policy and social policy. At this stage there were few reflections on the limitations of benchmarking. Since then some of the critical points made in this contribution have been taken into account.

In an important document from the Portuguese Presidency distributed at the end of the period (Council of the European Union, 9088/00, 14 June 2000) the use of benchmarking and its relationship to the open method of coordination was further discussed and clarified. In this document it is pointed out that the open method of coordination aims to organise a learning process about how to cope with common challenges of the global economy in a coordinated way while also respecting national diversity. Further it is specified that:

1. Benchmarking is only one element in the open method of coordination.
2. The open method is 'open' because best practises should be assessed and adapted in their national context, and because monitoring and evaluation should take the national context into account in a systemic approach.
3. The open method is also open in the sense that the method invites the participation of the various actors of civil society. It can be an important tool to improve transparency and democratic participation.

As pointed out in the introduction to the chapter the open method of coordination was introduced with two purposes. One was to promote European convergence also in areas where the legal foundation for European political integration is weak and the second was to stimulate institutional innovation in order to make Europe better prepared for the knowledge economy (for a more extended discussion of the new approach see Telo in this volume). It is too early to conclude on the outcome in these two respects but the European Commission documents referring to benchmarking give some idea about the potential for reaching these objectives.

IS THE NEW CONTEXT TAKEN INTO ACCOUNT IN EUROPEAN BENCHMARKING?

There seems to have been a wide acceptance at the European Summit among prime ministers that there is a need for institutional innovation in order to cope with the new context. In some policy areas, such as education and innovation policy, the documents from the European Commission presented at the summit in April were characterised by a clear understanding of what should be the implications of the movement into a learning economy. For instance it was emphasised that schools need to become learning centres and that new initiatives are needed to establish life-long learning.

In other areas, such as labour market and competition policy, there are long traditions for benchmarking on the basis of allocative efficiency with little reference to the impact on learning and innovation. Here the benchmarking exercises proposed by the European Commission seemed to reproduce the old patterns without too much thought about what the implications were of the new context. For instance how the institutional set-up of the labour market supports competence building was not taken sufficiently into account as compared to traditional measures of 'flexibility'.

A similar problem was seen in relation to competition policy where the performance of European firms will depend as much upon their positioning in global, national and local networks as upon the character of competition. Here old indicators of competition were motivated also by reference to their relevance for innovation but this appeared somewhat as an afterthought.

Looking at the follow-ups of these documents (Communication from the Commission, Structural Indicators, 25 September 2000) there has been some progress in these respects but the full implications of the co-existence of different national employment systems and different national patterns of industrial dynamics are not yet reflected in the sets of indicators used.

One major problem is to integrate the sectoral benchmarking exercises and the policy considerations they give rise to into coherent policy strategies at the regional, national and European level. In the context of the

Portuguese Presidency we have proposed the establishment of new high-level fora at the local, national and European level in charge of innovation and competence building to have a permanent responsibility for coordinating different policy areas. Here the June document from the Portuguese Presidency points to a stronger role for the European Council and an annual Spring meeting for prime ministers to discuss social and economic issues on the basis of an annual synthesis report to be produced by the Commission.

Innovation policy, systemic contexts and the US as 'best practise' new economy

In the recent documents covering innovation policy (Communication from the Commission to the Council and the European Parliament, COM (2000) 567 final) there is still some tension between the recognition of the co-existence of specific national innovation systems (p. 8) and the frequent references to the need to diffuse 'best practise' among member countries. A more explicit reflection on how to reconcile the two sets of ideas in operational terms would have been helpful. At the same time there is substantial progress in mapping the critical elements that constitute a performant innovation system.

Some of the small national systems (Finland, Netherlands, Denmark and Ireland) seem to be most successful in terms of innovation according to EU's preliminary attempts to establish an innovation scoreboard. It is interesting to note that this result coincides with the preliminary OECD analysis of the new economy, where some of these small countries appear together with the US as the most successful ones in coping with the new context. Since the Lisbon European Council, changes within the US raise some doubts about the sustainability of the US model of development.

In spite of this there is a tendency in the innovation benchmarking report to point to the US as the leader and to Europe as the laggard. Quite dubious *ad hoc* arguments are used to diminish the importance of the good performance of small European countries. Analysing the systemic features of the small European and welfare-oriented countries may actually be more fruitful than attempts to catch up with the US.

In the central documents from the period of the Portuguese presidency there are references to the objective to combine competitiveness with social cohesion. The possibility that the case of the small welfare economies demonstrate that there is a type of 'new economy', completely different from the US, that actually gets its strength from giving citizens security in change and from building social capital by sharing the costs of change is not explicitly considered, however. Of course, such a perspective raises some questions about how to pursue the European integration process without undermining social capital in these small countries and about how to build a similar type of social capital in the less homogenous bigger European countries.

This example demonstrates the limitations of using the benchmarking of separate policy areas when it comes to design coherent policy strategies. In order to understand what lies behind the relative success of some of the small countries it is necessary to see how social and cultural dimensions are co-evolving with organisational and techno-economic developments.

CONCLUDING REMARKS

There has been substantial progress in the understanding of the limits of benchmarking from the first documents that were produced for the Portuguese Presidency. The progress has been methodological as well as one of content. There is now a greater awareness in the texts coming from the Commission that 'best practise' needs to be considered relative to its national context. And the selection of indicators to be monitored have become more adequate when it comes to reflect the globalising learning economy.

What remains to be seen is if the ideals of bottom–up and democratic procedures announced in the June communication from the Portuguese Presidency can be realised. There is still a certain bias toward centralised and bureaucratic procedures to overcome. Another issue is how to integrate the sectoral efforts into coherent regional, national and European strategies. As it stands now the annual meeting of the Prime Ministers is expected to contribute to such an integration. But is that sufficient?

NOTES

1. The concept of 'the learning economy' was introduced in Lundvall and Johnson (1994) and it has been further developed in Lundvall (1998) and Lundvall and Barras (1998). The basic idea is that there has been a speed-up both in the rate of formation and destruction of knowledge and what really matters for economic success of individuals and organisations is the capability to learn (and forget).
2. There are other forms of post-modern management techniques getting into high fashion. One of them is the idea of using 'story-telling' in order to promote new ways of doing things or in order to create an organisation-wide common interpretation of reality. While benchmarking emphasises generic quantitative indicators, 'story-telling' emphasises specific qualitative dimensions. But the two techniques may overlap in the sense that some story-telling may be required in order to establish a common understanding of what is 'best practice'.
3. What is really existing best practice in a specific area might be far from the best possible practice. Comparing different industrial sectors in terms of their use of advanced organisational techniques in Denmark it was found that firms in the construction industry were

far behind firms in other sectors. This example illustrates that the most positive learning effects may come about when the benchmark is chosen outside the sector or by comparing a country with countries belonging to a different league or region.

4. Benchmarking may seen as an economic intelligence activity and the borderline may sometimes be unclear between non-cooperative benchmarking industrial espionage.

5. More recent contributions are Lundvall (ed.) (1992), Nelson (ed.) (1993), Edquist (ed.) (1997) and Amable, Barré and Boyer (1997). Porter (1990) did not explicitly use the concept but his approach is in important respects similar to the one found in the other contributions. McKelvey (1991) compares some of the early contributions while Freeman (1995) puts the concept into a historical perspective linking it to the works of Friedrich List.

6. On the one hand, the pure market could not carry the necessary qualitative information between users and producers. On the other hand, the transaction cost response: the transformation of the market into a hierarchy – was not forthcoming in the real world; empirical studies demonstrate that a majority of major innovations are product and not process innovations. The conclusion that coordination and cooperation between firms is a third alternative to market/hierarchy is consistent with the analysis of Richardson (1972), but based on a different analytical argument. While the emphasis of the latter is on coordination of activities that are similar and/or complementary, our emphasis was on the prerequisites for product innovation.

7. Both the World Bank and the OECD have recently intensified analytical work on 'social capital'. The OECD work is part of the on-going major OECD study on the new economy' and it will be published in the spring 2001.

8. The general emphasis on deregulation and on 'flexibility' in the OECD Job's Study as the major objective of labour market reform has been problematised by a series of recent studies (see Lindley in this volume).

REFERENCES

Allen, P.M. (1988), 'Evolution, innovation and economics', in G. Dosi, C. Freeman, R.R. Nelson, G. Silverberg and L. Soete (eds), *Technology and Economic Theory*, London: Pinter Publishers.

Amable, B., R. Barré and R. Boyer (1997), *Les systèmes d'innovation à l'ère de la globalisation*, Paris: Economica.

Bogan, C. and M.J. English (1994), *Benchmarking for Best Practices: Winning through Innovative Adaptation*, New York: McGraw-Hill.

Boxwell, R.J. Jr. (1994), *Benchmarking for Competitive Advantage*, New York: McGraw-Hill.

Cabinet Office (2000), Press release, CAB 38/00, 3 February 2000.

Camp R.C. (1989), 'Benchmarking: the search for best practices that lead to superior performance. Part I: Benchmarking defined', *Quality Progress*, 22(1), 61–8.

Coleman, J. (1988), 'Social capital in the creation of human capital', *American Journal of Sociology*, 94(Supplement), 95–120.

Commission of the European Communities (2000), Communications from the Commission to the council: Structural indicators for the Synthesis Report, Brussels, 29.09.00, Com (2000) S 94.

Commission of the European Communities (2000), Communications from the Commission to the council, the European Parliament, the Economic and Social Committee and the Committee of the Regions: Innovation in a knowledge-driven economy, 20.09.00, Com (2000) S67 final.

Council at the European Union (2000), follow-up of the Lisbon European Council

– the on-going experience of the open method of co-ordination, Brussels, 14.06.00, 9088100.

Coriat, B. (2001), 'Organisational innovation in European firms', D. Archibugi and B.-Å. Lundvall (eds), *The Globalising Learning Economy*, Oxford, Oxford University Press.

Cox J. R.W., L. Mann and D. Samson (1997), 'Benchmarking as a Mixed Metaphor: Disentangling Assumptions of Competition and Collaboration', *Journal of Management Studies*, 34(2), 285–314.

Dore, R. (1986), *Flexible Rigidities: Industrial Policy and Structural Adjustment in the Japanese Economy 1970–1980*, London: Athlone Press.

Dosi, G., C. Freeman, R.R. Nelson, G. Silverberg and L. Soete (eds) (1988), *Technology and Economic Theory*, London: Pinter Publishers.

DTI (1998), *Closing the Gap: The Performance of UK SMEs within the United Kingdom Benchmarking Index*, London: Department of Trade and Industry.

Edquist, C. (ed.) (1997), *Systems of Innovation: Technologies, Institutions and Organisations*, London: Pinter Publishers.

European Commission (1998), 'Benchmarking skills: final report', EC-Directorate General III – industry, and Ministry of Industry and Energy, Spain.

European Commission (1999), 'First report by the high level group on benchmarking. Benchmarking Papers No.2', EC-Directorate General III – industry.

Freeman, C. (1987), *Technology Policy and Economic Performance: Lessons from Japan*, London: Pinter Publishers.

Freeman, C. (1988), 'Japan: New national innovation systems?', in G. Dosi, C. Freeman, R.R. Nelson, G. Silverberg and L. Soete (eds), *Technology and Economic Theory*, London: Pinter Publishers.

Freeman, C. (1995), 'The national innovation systems in historical perspective', in *Cambridge Journal of Economics*, 19 (1) 5–24.

Garvin, D.A. (1993), 'Building a learning organisation', *Harvard Business Review*, 71(4), 78–91.

Green, F., S. McIntosh and A. Vignoles (1999), '"Overeducation" and skills – clarifying the concepts', Centre for Economic Performance, LSE, Discussion Paper dp0435, ISBN 0 7530 1306 1.

Kline, S. J. and N. Rosenberg (1986), 'An overview of innovation', in R. Landau and N. Rosenberg (eds), *The Positive Sum Game*, Washington DC: National Academy Press.

Kutznets, S. (1960), 'Economic growth of small nations', in E.A.G. Robinson (ed.), *Economic Consequences of the Size of Nations*, Proceedings of a Conference held by the International Economic Association, London: Macmillan.

Leibfried, K.H.J. and C.J. McNair (1992), *Benchmarking: A Tool for Continuous Improvement*, New York: Harper Business.

List, F. (1841), *Das Nationale System der Politischen Ökonomie*, Basel: Kyklos (translated and published under the title *The National System of Political Economy* by Longmans, Green and Co., London 1841).

Lundvall, B.-Å. (1985), *Product Innovation and User–Producer Interaction*, Aalborg: Aalborg University Press.

Lundvall, B.-Å. (1988), 'Innovation as an interactive process: From user–producer interaction to the national innovation systems', in G. Dosi, C. Freeman R.R. Nelson, G. Silverberg and L. Soete (eds), *Technology and Economic Theory*, London: Pinter Publishers.

Lundvall, B.-Å. (ed.) (1992), *National Innovation Systems: Towards a Theory of Innovation and Interactive Learning*, London: Pinter Publishers.

Lundvall, B.-Å. and S. Barras (1998), *The Globalising Learning Economy: Implications for Innovation Policy*, DG XII-TSER, Brussels.

Lundvall, B.-Å and Johnson, B. (1994), 'The learning economy', *Journal of Industry Studies*, 1 (2, December), 23–42.

Lundvall, B.-Å. and P. Nielsen, (1992), 'Competition and transformation in the learning economy – illustrated by the Danish case', *Revue d'Economie Industrielle*, 88, 67–90.

Macneil, J., J. Testi, J. Cupples and M. Rimmer (1994), *Benchmarking Australia: Linking enterprises to world best practice*, Melbourne: Longman Business and Professional.

Marsden, D. (1999), *A Theory of Employment Systems: Micro-Foundations of Societal Diversity*, Somerset: Oxford University Press.

Maskell, P., H. Eskelinen, I. Hannibalsson, A. Malmberg and E. Vatne (1998), *Competitiveness, Localised Learning and Regional Development. Specialisation and Prosperity in Small Open Economies*, London: Routledge.

McKelvey, M. (1991), 'How do National Innovation Systems differ?: A critical analysis of Porter, Freeman, Lundvall and Nelson', in G. M. Hodgson and E. Screpanti (eds), *Rethinking Economics: Markets, Technology and Economic Evolution*, Aldershot: Edward Elgar.

Mosley, H. and A. Mayer (1998) 'Benchmarking national labour market performance: a radar chart approach, final report', Report prepared for European Commission, Directorate-General V, Employment, Industrial Relations and Social Affairs (DGV/A2), *Wissenschaftszentrum Berlin für Sozialforschung*.

Nelson, R.R. (1988), 'Institutions supporting technical change in the United States', in G. Dosi, C. Freeman, R.R. Nelson, G. Silverberg and L. Soete, (eds), *Technology and Economic Theory*, London: Pinter Publishers.

Nelson, R.R. (ed.) (1993), *National Innovation Systems: A Comparative Analysis*, Oxford, Oxford University Press.

OECD (1999), *OECD Science, Technology and Industry Scoreboard: Benchmarking the Knowledge-Based Economies*, Paris: OECD.

OECD (2000), *Knowledge Management in the Learning Society*, Paris: OECD.

Polanyi, M. (1958/1978), *Personal Knowledge*, London: Routledge & Kegan Paul.

Polanyi, M. (1966), *The Tacit Dimension*, London: Routledge & Kegan Paul.

Porter, M. (1990), *The Competitive Advantage of Nations*, London: Macmillan.

Putnam, R.D. (1993), *Making Democracy Work – Civic Traditions in Modern Italy*, Princeton, NJ: Princeton University Press.

Richardson, G. B. (1972), 'The organisation of industry', *Economic Journal*, 82, pp. 883–96.

Rothwell, R. (1977), 'The characteristics of successful innovators and technically progressive firms', *R&D Management*, 7(3), 191–206.

Sako, M. (1990), 'Buyer-supplier relationships and economic performance: evidence from Britain and Japan', Ph.D. thesis, University of London.

Spendolini, M.J. (1992), *The Benchmarking Book*, New York: AMACOM.

Watson, G.H. (1993), *Strategic Benchmarking: How to Rate Your Company's Performance Against the World's Best*, New York: John Wiley.

Whitley, R. (1996), 'The social construction of economic actors: institutions and types of firm in Europe and other market economies', in R. Whitley (ed.), *The Changing European Firm*, London: Routledge.

Woolcock, M. (1998), 'Social capital and economic development: toward a theoretical synthesis and policy framework', *Theory and Society*, 27(2), 151–207.

7. The construction of European identity

Manuel Castells

PRELIMINARY REMARKS

This is not a research paper. It is a policy-oriented document prepared at the request of the Portuguese presidency of the European Union in the first semester of 2000. It does rely on a number of materials, information, and analyses, most of which can be found in my trilogy 'The Information Age: Economy, Society, and Culture' (Oxford: Blackwell, revised edition, 2000). In this context, I will not try to demonstrate my points. This is written in statement mode, to suggest ideas and directions for discussion, with the purpose of advancing in the construction of European identity. I will not be dealing here with major strategic issues, such as economic policy, technological development and the restructuring of the welfare state, without whose proper treatment, discussions about identity become an empty ideological exercise. I am aware of other contributions included in the same volume, prepared for the Portuguese presidency of the European Union on these matters. They are excellent papers, I agree with them in their main lines, and I will avoid to be redundant with their contribution. I will start with one previous question: is European identity a relevant matter in the construction of the European Union?

1. WHY EUROPEAN IDENTITY IS IMPORTANT

After the creation of the euro, and the constitution of the European Central Bank, the European Union is, for all practical matters, one economy – waiting for the full integration of UK and Sweden to consolidate the union. Any reversal in the process of integration in the coming years would have catastrophic consequences for European economies, and for the global economy. Besides the economic dimension, European Union countries are now intertwined in a web of institutional, social and political relationships which will grow in size and complexity in the coming years, as new countries become associated with the EU, and as the European institutions extend their realm of activity. Thus, we are too far into the process of European integration (with considerable benefits for everybody, at this point) to think the unthinkable: the future breakup of the European Union. And yet, the European ground may be shakier than we believe. This is, first of all, because the global economy is, and will be characterised, by recurrent crises, in the financial markets, in trade arrangements and in the integration of social, national, cultural and environmental demands from people around the world – as WTO's fiasco in Seattle has shown. We were lucky in the 1997–9 crises in Asia, Russia and Brazil, that financial turbulences were contained within emergent markets. But as core markets become electronically entangled worldwide, and as the dynamism of the new economy is coupled with high doses of risk and unpredictability, we cannot bet on a smooth transition to full-fledged globalisation. Besides, the transition to a new technological paradigm, and to a new economy, that is only now picking up speed in Europe, is bringing substantial disruption to important segments of the population, in many regions, and is affecting the interests of social actors and political institutions which were rooted in a very different economy and society. If we add to this, the increasing multi-ethnic character of most European countries, and the emergence of new kinds of geopolitical dangers (nationalism and fundamentalism from the excluded and marginalised of the new economy), I think it is fair to say that we are heading toward a very stormy period in spite of the extraordinary potential in an age of creativity, prosperity, and institutional reform. It can go both ways depending on what Europeans do. The 'Europeans' are of course the tricky part of the equation. Because, who are they?

As long as the European Union is a positive sum game, in which everybody wins (some in economic terms, some in political terms, others in technological terms, still others in social terms), without sacrificing too much national identity and political sovereignty, crises of transition are absorbed by the countries themselves. Yes, the European Commission is not very popular, and its pitiful performance in 1997–9 has made things worse, although the first months of the Prodi regime have reinvigorated the Commission and given some hope for its future. But, in spite of their distrust of the Commission, people around Europe did not feel (rightly) that the Eurocrats had real power over their lives. Things are changing. Regardless of how much real power Brussels has or will have, the European Union as such, and other supra-national institutions (such as NATO, WTO or IMF), have taken away substantial areas of sovereignty from the European states. Not that nation-states are disappearing. But they have become nodes, albeit decisive, in a broader network of political institutions:

national regional, local, non-governmental, co-national and international. Europe is already governed by a network state of shared sovereignty and multiple levels and instances of negotiated decision making.

Thus, on the one hand, we are heading towards a complex process of economic/technological/cultural transition that will create innumerable problems and resistances – along with new opportunities and wealth. On the other hand, the political system in charge of managing the transition is increasingly disjointed from the social and cultural roots on which European societies are based. In other words: the technology is new; the economy is global; the state is a European network, in negotiation with other international actors; while people's identity is national, or even local and regional in certain cases. In a democratic society, this kind of structural, cognitive dissonance may be unsustainable. While integrating Europe without sharing an European identity is a workable proposition when everything goes well, any major crisis, in Europe or in a given country, may trigger a European implosion of unpredictable consequences. Because the construction of identity is a long-term process, we are in a race against the clock between the time horizon of transitional, social/economic crises and the emergence of a European identity on whose behalf citizens around Europe would be ready to share problems and build common solutions. Instead of blaming the neighbour, and de-legitimising their governments, potentially suspect of eurocracy.

2.   WHAT IS EUROPEAN IDENTITY?

For the sake of clarity, identity is a set of values that provide symbolic meaning to people's life by enhancing their individuation (or self-definition) and their feeling of belonging. Or course people may have various identities, according to different spheres of their existence: one can feel Portuguese, socialist, catholic, female, and all these identities can overlap without major contradiction. Which one is dominant depends on the moment in life and on the realm of activity.

European identity would be the set of values that would provide shared meaning to most European citizens by making it possible for them to feel that they belong to a distinctive European culture and institutional system that appeals to them as legitimate and worthwhile. Which could be the sources of such an identity?

It is essential to know, first, what is *not* European identity. It is not a 'civilisation' based on religion, past history or a set of supposedly superior 'Western values' (a la Hungtinton). European countries have spent centuries (and particularly the last one) killing each other, so the notion of a shared history has a sinister connotation. Religion (meaning Christianity) is an unthinkable source of identity once we have established the separation between the church and the state and at a time when non-Christian religions (e.g., Islam) are growing fast in the European Union, both among ethnic minorities, and in future member countries (Turkey). Language, one of the most important sources of cultural identity, is, of course, excluded as a common source of European identity, although I will argue that a certain approach to language is essential in constructing identity. National identity as a European identity is also impossible, by definition. Nations and nation-states are not going to fade away. In fact, they are going to grow and become important sources of collective identity, more than ever, as new, formerly oppressed nations, come into the open (Catalunya, Euzkadi, Galicia, Scotland, Wales, Wallonie, Flanders, etc.), and as strong nationalist movements assert their rights in public opinion against the submission of the nation to the European state. I start from the assumption that in the foreseeable future, Europe will not be a federal construction similar to the United States. There will be no unified European state, superseding and cancelling current nation-states. Identification with a political construction, such as the state, cannot be a source of identity, thus eliminating the option of 'European nationalism' equivalent to 'American nationalism'. American national identity emerged from a multicultural, immigrant nation. But it was because it was an immigrant nation in an empty continent (or forcefully emptied of its native inhabitants when necessary) that America could combine strong cultural and ethnic identities with an equally strong American identity. Such is not, and will not be, the case for Europe.

So, it is in the realm of values, of new values where we could find the seeds of a European identity. On the basis of surveys of attitudes, and a review of the literature, in my book *End of Millennium* (2000), I identified some elements of what I called a 'European identity project'. This is not what I propose, but what appears empirically to carry a broad cultural consensus throughout Europe, besides the values of political democracy (which is a widely shared value, but not distinctively European). These elements can be identified as shared feelings concerning the need for universal social protection of living conditions, social solidarity, stable employment, workers' rights, universal human rights, concern about poor people around the world, extension of democracy to regional and local levels, with a renewed emphasis in citizen participation, the defence of historically rooted cultures, often expressed in linguistic terms; for women, and for some men, gender equality. If European institutions were able to promote these values, and to accord life and policy with these promises for all Europeans, probably this 'project identity' would grow. But the problem is precisely that some of these aspirations will have to be re-thought and adapted in the new historical context, for instance in what concerns the welfare state or stable employment. Moreover, the mere enumeration of these values shows that, while they are a reasonable wish list, it may not be easy to combine them in a coherent set, beyond their popularity in public opinion. So, these elements of a European project, while they are materials to work with, cannot be asserted as a finished model to be imposed top–down as, for instance, the French revolution did with its political ideals, to construct, at the same time, the universal citizen and the French nation, as necessary and sufficient conditions of the civilised state and society. This extraordinary accomplishment could be carried on only through military and political domination, and under conditions of restrictive democracy (without women's voting rights, and without the tolerance of historic cultures beyond the Ile-de-France). In a fully democratic, multicultural, multiethnic Europe, exposed to global flows of communication and information, no project can be imposed from the state. Thus, European identity does not exist, and there is no model that could be taught and diffused from the European institutions and national governments.

Yet, the problems I raised are still relevant. While national and local identities will continue to be strong and instrumental, if there is no development of a compatible European identity, a purely instrumental Europe will remain a very fragile construction, whose possible, future wrecking would trigger major crises for European societies.

If there is no European identity model, there still can be an identity in the making, that is a process of social production of identity. In other words, it is not possible to create, artificially, a European identity, from a 'concours d'idées' in the same way that, at one point, the Yeltsin government was trying to find a new Russian identity. However, European institutions could help in the development of a series of mechanisms that, in their own dynamics, would configurate the embryos of this shared system of values throughout Europe. European governments should also set up a system of observation capable of detecting the birth and development of these new values, and to ensure their diffusion, and interaction, while avoiding transforming them into a new ideology, the ideology of pan-Europeanism in this case. It is by engaging in social experimentation, by letting society evolve by itself, but helping to construct a European civil society, that we could see the emergence of a new European identity in a few years from now. The description of some of the potential processes that could induce such identity will help to make this discussion a matter of concrete policy.

3. QUESTIONS OF METHOD: BUILDING IDENTITY BY MAKING SOCIETY

First, who are the actors of this identity construction process? Let me be clear, in the current state of affairs they are mainly the European national governments, acting through the council of ministers of the European Union. The Commission can only be a relatively autonomous manager of shared political decision making. Any attempt to make the Commission the center of power and sponsor of new identity will ultimately provoke the revolt of national and local identities, thus jeopardising the European Union. The European Union is not, and will not be, a classical federal state. It is a new form of state. And in this new form of state the connection to societies rests on the various nodes that assume direct political representation. The construction of a European identity, if it ever happens, will be the fact of European societies, under the strategic impulse of the Council of Ministers, reflecting a common project shared across the political spectrum by the countries participating in it. To say so is to say that there cannot be an agreement on the content of European identity (for instance, between conservatives and socialists, between ultra-nationalists and greens, etc.) But a consensus could be built on the method, on the mobilisation of societies towards new, shared values that would be widely diffused throughout Europe, so that every party, interest group, or ideology, would hope to win in the process: it is similar to sharing democracy as a method, without having to agree on the substance of politics. Democracy, besides being a principle, is, in practical terms, a method, of political representation and governance.

What could be the method of identity building, shared between countries and political forces throughout Europe? Here is where this contribution must dare to become speculative and prospective, since these are entirely uncharted waters.

4. SOCIAL AND INSTITUTIONAL PROCESSES OF EUROPEAN IDENTITY BUILDING

Remember: all processes of production of identity are based on a common methodology. Identity is built by sharing cultural and social practices throughout Europe, letting the outcome of this sharing emerge from the experience. In other words, we do not know what this European identity will be, but we could create the material possibilities for its emergence from society, then reinforcing and communicating the developing identity with the help of European institutions. I certainly know that some of the elements of the mechanisms proposed are already in practice in Europe. I am simply emphasising them, and adding other proposals which are less diffused or non-existent.

The first, and most obvious, of such mechanisms is *education*. There must be introduced at all levels of the education system of every country, some common elements, including the history, and culture, and language of other countries in the programs of all schools. Hopefully larger proportions of teachers and students will spend periods of their school activity in other countries, along the lines pioneered by the Erasmus program for university students. Yet, a true interpenetration of education systems requires a serious effort, and a concerted policy of European countries in this direction. Equivalence of pedagogic systems and programs (which does not mean making uniform all programs, quite the contrary) will allow passages from one system to another, and will make it possible to use the degrees obtained in one country over the entire European market, in real terms (current possibility is only on paper, since in most cases, qualifications and language skills are not really equivalent, from the perspective of employers. Indeed: less than 3 per cent of European Union citizens work in an EU country different from their own country).

The second mechanism, still to be explored, is the widespread diffusion and use of Internet in the population at large. Internet is a privileged tool of communication and access to information. It is not just a technology, it's the economic, cultural, and political backbone of the information age. An Internet literacy campaign, aimed mainly at the adult population (the children will have it at school) would provide the communication bridge for Europeans, and would bring all societies to the same technological level. The model experience of Finland in this sense could be an example from where to build.

The third mechanism is a pan-European linguistic policy, aiming at the cross-cultural diffusion of all languages in all countries, through the education system at all levels, via internet, by cultural programs, etc. I am always chagrined when I see how in American leading universities students can learn not only major languages, such as Portuguese or French, but also Catalan, Finnish or Swahili. True, most American students do not learn languages. But, with cultural and educational incentives, European students would. And this multilinguistic web could be a source of true multiculturalism.

Fourth, we need a pan-European media policy. The coming of multimedia may be dominated by Hollywood and San Francisco and New York multimedia designers, and by global mega-conglomerates, such as the one prefigurated by AOL/Time-Warner. The European reaction is nationalistic, defensive and ultimately doomed in a market economy. In the age of internet and satellite communication it is not by imposing quotas that we will bring people to alternative sources of culture and communication, different from the ones currently dominating in the business world. Europe should not subsidise private groups just because they are European, but European governments should allow their merger and strengthening, or they will not be competitive. European governments should also act in favouring the development of high-quality, competitive, publicly subsidised, multimedia groups operating independently under a charter. Modelled upon the performance of the BBC, an independent, high-quality, globally competitive, multimedia group, a network of joint ventures among public European TVs and studios should develop. It should also open up, from the beginning to private Internet Service Providers, in order to position European cultural senders in the coming process of technological convergence in the media system. A pan-European media system, both public and private, will be the cornerstone of an emergent European culture.

Not everything is culture: geographical mobility of labour is essential to build a common European experience. The conditions under which southern Europeans emigrated to northern Europe must not be repeated. The integration of a labour market would require access to housing and social services, equivalent professional qualifications and equal rights. If people truly can work everywhere in Europe, not only will the economy reap extraordinary benefits, and unemployment be reabsorbed, but people will experience in real terms other life styles, other cultures. If this is accomplished in conditions of equality and non-discrimination, the Europe that works together will learn to live together.

On the condition that we tackle initially the issue of multiethnicity and multiculturalism. Europe is fast becoming a continent of ethnic minorities. The proportion of foreign-born population in Germany is already almost the same as the African-American population in the US, at about 12 per cent. And, as for African-Americans, most people from ethnic minorities concentrate in the largest metropolitan areas, thus increasing their visibility. Because of the differential birth rate *vis à vis* native populations, the coming two decades will bring a spectacular increase of multiethnicity throughout Europe. If we add the future integration of Eastern European and Turks into the European Union, Europe must design from now on specific policies of cultural integration, based on equal rights, and respect for differences, that should be applied throughout the continent. In addition, Europe needs a new immigration policy that could attract the necessary talent that exists around the world, and would be open to genuine political refugees, but, at the same time should clamp down on illegal immigration, and particularly on the mafias that are bringing into Europe about half a million undocumented immigrants every year. In addition, policies of easy naturalisation for lawful residents should be designed, and applied in similar terms in all countries. The building of an European identity can only proceed on the basis of the acceptance of its multiethnic, multireligious and multicultural character, and this acceptance needs a material basis in immigration and naturalisation policies, in multiculturalism in the education system, and in the openness of the media and of cultural institutions to the diversity of cultural expressions.

Building bridges in Europe means also building bridges between European cities and between European regions. There are already a number of dynamic networks and dynamic institutions, including the Committee of Regions, an advisory body to the European Commission. National governments should accept, and encourage the initiative of sub-national governments to establish their own European networks. A defensive attitude from national governments in this matter will lead to endless internal conflicts, while cities from different countries will be competitors rather than cooperators. Inter-municipal, and inter-regional European networks are essential sources of reconstructing culture, besides yielding considerable economic benefits.

Similar pan-European networks exist among business organisations, labour organisations, cultural associations (such as European artists) and citizen groups. With or without governmental support, the creation of this European network of social actors is another layer of identity construction, as they are the embryos of a European civil society.

More complex is the issue of political identity. It cannot be built through allegiancy to an unlikely European federal state. In this sense, the European democratic deficit does not come from the powerlessness of the European Parliament. The strengthening of the Parliament would lead to true supranationality and federalism, something that most public opinion, and most political parties would not tolerate. The European Commission does not have to be submitted to the European Parliament, but to the European governments, to the Council of Ministers. The key democratic issue is the transparency for citizens of what the Council of Ministers does, and the explicit inclusion of European policies in the political platforms of parties in the national, regional and local elections. European democracy is not accomplished by removing institutions from their roots of representation, but by bringing European institutions down to where citizens live and feel. However, increasing the activity and role of the European parliament, and connecting it more explicitly to a European constituency, is one element, among others, in the building of a shared identity.

Last but not least, European identity will be built around a common international policy, which includes a common defence policy. Only if citizens realise that by being Europeans they can act upon global issues in terms of their own values and interests, will they realise how important it is. Under current conditions, European international policy is confused, non-existent, or powerless. And, among other things, it is powerless because it is entirely subordinated to NATO in terms of defence. Europe needs independent European armed forces, with full technological and operational capability, working in close cooperation with NATO and with the United States. But to be able to assert this independence, Europe needs to invest in technology, to increase its military budget and to train multinational, professional armed forces. It is hypocritical to resent 'American hegemony', and then call upon the United States each time there is a serious security crisis, letting the US foot the bill in resources, and personnel. If Europe wants autonomy, it has to assume its fair share of the Western defence burden. The recently created Rapid Deployment Force is a good beginning, but too modest and without the actual military capability for acting independently. To assert itself in the international arena, as a unit, would clarify for Europeans the values and strategic goals for which Europe stands. But for this not to be empty rhetoric, and able to permeate down the consciousness of citizens, European governments would need to set up a common system of international representation, coordinating their presence in international institutions, and provide Europe with the financial and military means to back up its positions. Only then could emerge, truly, a European political identity, as one of the dimensions of European identity.

Finally, since all these mechanisms are very indirect approaches to the building of a European identity, European institutions should be able to monitor their development and to identify the actual elements of European identity as they emerge from practice in society. So an European Identity Observatory should be constituted, based on a network of observers and analysts, with a very light infrastructure, very economical means, no power, and as independent as possible. It should report on an annual basis on the level of development of European identity, and on the substantive elements that appear to configurate the emerging model. European institutions could start modelling themselves according to the cultural expressions and organisational forms emerging from civil societies throughout Europe. Maybe then new forms of democracy could emerge, as states learn to follow and adapt to the evolution of society.

8. Governance and government in the European Union: The open method of coordination

Mario Telò

1. THE GLOBALISED WORLD, NEW REGIONALISM AND THE EU BETWEEN CONVERGENCE AND DIVERGENCE

Despite growing interdependence and enhanced pressures towards a common pattern, it is not entirely possible to eliminate socioeconomic divergence and capitalist diversity within the EU. Not only do they have deep historic, cultural, social and economic roots, but, more importantly, it has become clear for international literature that even though the globalisation and integration process pushes for a greater convergence, it goes paradoxically together with a deepening of 'localisation', that is a deepening of regional and national differences.[1] James Rosenau has rightly proposed the concept or neologism of 'fragmigration' to grasp this mixture of integration and fragmentation at the level of the global system.[2] This new reality is bound to have consequences on the elaboration of international governance norms and on their chances of being consistently implemented. The international coordination of policies, especially at the Trilateral Commission level, was a tentative response at the end of the long period of US-centered hegemonic stability, which emerged after the Second World War, during the decades of embedded capitalism. Its success, mitigated during the 1980s, proved to be increasingly insignificant over the last decade of the century; this implies a risk of exacerbating discord economic globalisation, and also social competition among national and sub-national systems (but with a rush to the bottom[3]) and instability, both at the global level and within national societies.

This chapter places the systemic development of neo-regionalism, and more specifically of the European Union, in the framework of the growing gap between the supply and demand of good international governance. National, world and regional organisations and institutions will influence international modes of competition in the market share, the social cohesion and the political stability in the world. The EU is paradoxically perceived as both a deregulating and as a 'reregulating' actor (as shown by the Danish referendum in September and the demonstrations in December 2000 in Nice for example).

In certain conditions, neo-regionalism can become a particularly adequate, efficient and legitimate level of governance of modernisation, within the framework of a multilevel international governance and as a support to national and sub-national governance.

As far as its contribution to world governance is concerned, neo-regionalism has become a structural factor of the post-Cold War globalised international system; it can contribute to filling the good governance deficit through a reduction of the asymmetry of the post-Cold War system and also through resolution of the problems caused by the fragmentation and excessive number of actors. Despite the heterogeneity of regional organisations and arrangements in the current global system (ranging from soft to deep integration), in general regional organisations already go beyond the simple economic and commercial dimension and constitute a kind of third level between the global level, increasingly perceived as powerless and lacking legitimacy, and the state level, affected by the loss of external and internal sovereignty.

The European Union is the key region in the perspective of a more symmetric and less-fragmented global governance: although unique in terms of its institutional features and of its history, as a laboratory, it is one of the main reference for the course of the other regional organisations (Mercosur, Asean, Sadc, Ecowas, Comunidad Andina, etc.), and is a source of emulation. More importantly, it acts consciously as a proactive factor of the regionalisation of the world through its external policies: common commercial policy, cooperation policy, inter-regional agreements, including the political dialogue. Of course, the European model is specific not exportable. But it can play a crucial catalyst role in one aspect: its message must be clear, both at the level of content[4] and the mode of governance of the internal diversities. It must in some way be translatable into the experiences of other regions of the world, concerning their methods and institutional forms of integration.

The 'Lisbon strategy' has been innovative at these two levels: as far as content is concerned, it has put a model for a knowledge-based society and for socioeconomic regional modernisation for the long term on the agenda of European states; this constitutes the soul of neo-regionalism, without which globalisation will turn into the assertion of a unilateral globalism, in the sense of a unipolar and asymmetric strategy which will only worsen conflicts and weaken global governance. As far as regional institutions are concerned, the 'Lisbon strategy' has opened up a new track, which is likely to produce increased socioeconomic integration, while at the same time is more easily translatable into the language of the other regional organisations because it is more capable of conciliating further integration showing respect for national diversities. This chapter will deal more particularly with this aspect, the new methods for the reorganisation of governance and internal government of the EU, with a view to strengthening convergence at regional supranational level.

With regard to internal institutional development, in Western Europe, convergence has been considered a common goal since the start of European construction, in spite of national differences (see Preamble to the Treaty of Rome). The multiple successes of the European construction, its deepening and its successive enlargements (from six founding States to the current fifteen and the twenty-seven predicted at the end of the enlargement process launched in Helsinki in 1999) seem to confirm this route. Three values were at the basis of convergence during the first decades of European construction: peace between former enemies and stability; national and supranational democratisation; social and economic prosperity. Henceforth secured in Western Europe, since 1989, the first two have largely been considered important mobilising factors at the level of the continent, coming as a justification for Eastern enlargement, despite its risks and implications. More specifically, for the first time in European history, peace and democracy are at the core of a rational and credible continental project. The real problem is that the conditions for a positive trade-off between enlargement and deepening are, on the one hand, the success of the institutional reform, launched in Maastricht and Amsterdam, which went through so many difficulties during the Nice IGC, and, on the other, the successful modernisation of the European socioeconomic model.

Thus, the role of the EU both as a global actor and as a continental driving force depends on its capacity to successfully modernise the economic and social model. And therein lies the problem: it is highly controversial among the fifteen European states and public opinion varies as to whether economic and social prosperity, that is conciliation between economy and society, between global competitiveness and social justice, should justify an increased European integration, particularly if traditionally conceived as a 'new stage in the process of European integration',[5] towards 'an even closer Union among the peoples of Europe'.[6] There is no unanimity, either at the national level of public opinion, the elite, or of the scientific community in terms of the kind of dynamic rebalancing to be pursued between the different levels of European, national, sub-national and supranational governance.

2. THE SEARCH FOR NEW METHODS OF EUROPEAN INTEGRATION

The 'Lisbon strategy' is taking up two major challenges: to identify the new undertakings that have to be launched in order for Europe to recover from the delay in the construction of a knowledge-based society, and above all to establish how to launch, manage and apply the innovation programme. The question at the core of these challenges is how to organise the governance of the economic and social modernisation of the European region. The Lisbon European Council, whilst recognising that national and regional differences are of value and that it is not possible to forget them, has stated that the current drift towards 'competition states'[7] has the negative effect of increasing intra-European tensions, generating defensive and/or fragmented approaches to globalisation, thus favouring the socioeconomic decline of Europe as a whole within the globalised world. The 'How?' thus becomes the key challenge. It is precisely at the level of governance and of reorganisation of the balances of power between the four European institutions (Council, European Council, Parliament and Commission) that EU has achieved limited results, despite the scope and quality of the projects relating to similar objectives, namely the harmonisation projects of the 1970s and 1980s (the Social Action Programme of 1972) and, more recently, the famous White Book of the Delors Commission in 1993.[8]

The starting point should therefore be the difficulty for the European Union to be proactive in these areas, crucial to its future, a difficulty which recently almost led to the blocking of the political initiative. Traditionally, the strategies aimed at reducing the differences between the member states have called upon two main methods: intergovernmental cooperation and community integration. As far as the first method is concerned, several scientific works have confirmed the crucial importance of national mobilisation of interests in favour of Europe and the weight of national preferences for Europe expressed by governments, in short, the European choices of states.[9] However, intergovernmental cooperation as such has proven to be dramatically inadequate during the last decade, and not only with regard to the two political pillars of the TEU (II and III).

Secondly, the federal-functionalist method has delivered extraordinary results for fifty years, paving the way to monetary Union and to the creation of a *sui generis* political entity: the EC/EU. Nonetheless, the international scientific literature largely agrees on the fact that there has been a turning point since the end of the 1980s, following internal and external factors directly affecting the federal-functionalist method: the politicisation of the issues of integration at the heart of the majority of member states and the emergence of acute problems of legitimation, not only towards new transfers of national competencies to the EU, but also towards the already existing 'normative supranationality'.[10] Moreover, the new international framework, that is the globalised world and the collapse of the communist regimes in Eastern Europe, has forced the European Union to deal simultaneously with two contradictory challenges. On the one hand, increased internal diversity: global competition and the Eastern and Mediterranean enlargement led to growing differentiation, complexity and internal heterogeneities. On the other hand, the need to restructure its governance system and to strengthen the central government. Growing world political responsibilities and the negative effects of purely national competitiveness raise, once again, the question of unity and convergence. To sum up, while both diversity and the need for convergence grow, the traditional integration method does not meet the internal consensus required to enable new delegations of national competencies from the states to the EU.

This historic turning point is basis of the eager search for new methods, which has characterised the EU, especially since the Treaty of Maastricht. The Treaty of European Union and the difficulties in its ratification are a sign both of the scope of the turning point underway and of the difficulty of this quest for a new European model. The 'opting-outs' are a huge change in Community law, since, for the first time, internal differentiation (ex: EMU) is based on non-will (with no time limits) of a state to take part in a common policy and not on its impossibility (transitory, and anyway, based on objective reasons), in accordance with the traditional model of a 'two-tier Europe' presented in the famous Tindemans Report of 1976. The multiplication of exceptions, derogations and extra-Treaty agreements and protocols (Schengen Treaty and Maastricht Social Protocol, for instance) have been a consequence of this institutional difficulty. The highly controversial debate on 'enhanced cooperation' is also to be situated in the framework of the feverish search for new methods, through and beyond the Treaties of Maastricht and Amsterdam: provisional differentiation within the shared process towards the Union, or flexible integration, meaning potentially permanent differentiations and therefore open recognition of the heterogeneity of the final goals? This still remains an open question in spite of the steps accomplished by the Nice Treaty. Anyway, the chances of enhanced cooperation being applied to the first pillar are very slim, and impossible for policies in areas affecting the internal market or competition. To sum up, the deadlock of the old methods is the basis of the search for a multiplicity of new tracks. The answer launched by the Lisbon European Council of March 2000 is set in this context: it is relevant with regard to multiple and crucial areas of economic and social modernisation.

The Treaty of Maastricht had a 'new beginning' character. However, it is characterised by a double internal asymmetry with regard to the methods and means of integration, particularly important because the areas at stake are very close to the modernisation of the European socioeconomic model: a) first asymmetry: within Economic and Monetary Union, the imbalance between the federal and centralised character of monetary union and the decentralised character of economic union (Article 99 TEC); b) second asymmetry: between EMU, on the one hand, and the social union, on the other.

Some comments have presented the opting-outs and the above mentioned asymmetries as temporary, by citing the example of the Social Protocol, integrated into the Amsterdam Treaty. Nevertheless, one decade after Maastricht, these two asymmetries cannot seriously be considered as a transitory stage leading necessarily to the extension to the economic, social and political areas of the procedures applied to monetary union. It should be borne in mind that the EU has not progressed in that direction and that the teleological expectations have been completely disappointed.

What is even more interesting is that integration has nonetheless made progress, the Single European Act and Monetary Union have called upon the dynamics of integration and cooperation, but following new tracks and directions. The EU would not have progressed at all if it had stuck to the two classic methods. For example, the asymmetry between EMU and social union was corrected – even if only very partially – by the Treaty of Amsterdam, with the inclusion of the Social Protocol in the body of the Treaty and the new Employment Title. However, it should be stressed that this was only possible thanks to a new method.[11] Moreover, with regard to the Social Chapter of the TEC (Title XI), there is the method of social dialogue between the main social actors prior to the European regulation. This is the agreement between the social partners that creates the conditions for the elaboration and approval of a directive (in the same way as is done at the national level in Scandinavia and Germany, in accordance with the model which Ph. C. Schmitter calls 'social neo-corporatism'). Secondly, new voluntarist methods of soft social governance have been put into practice, even though they were not expressly provided for in the Treaties: for example, what W. Streeck calls 'governance by persuasion' (through resorting to expertise) or 'governance by consultation' (European parliament, social dialogue) or even 'governance by diffusion of best practices'.[12] Thirdly, concerning the new Employment Title of the Treaty of Amsterdam, a new and very promising method was introduced in the Treaty: multilateral surveillance (articles 125–30 TEC) which paved the way to the Luxembourg European Council and the 'Lisbon strategy'.

This lead us to the central hypothesis of this chapter: the opening up to new methodologies in European socio-economic regulation is inevitable if the EU wants to progress further in the process of convergence without instigating the above-mentioned internal oppositions and external difficulties. In particular, the 'Lisbon strategy' could not proceed without the development of new methods and new tracks because it deals with topics for which, in general, there is no EU competence. These are, paradoxically, the most important and delicate topics in national agendas, while it is not at all likely that the national and subnational authorities or the socioeconomic actors are willing to strengthen a system of community rules at the European level on these matters.

3. THE NEW METHODS IN THE FRAMEWORK OF THE EUROPEAN MULTILEVEL GOVERNANCE

The 'Lisbon strategy' must be analysed within the framework of the vision of globalisation which is its prerequisite. Reading of the documents from the Presidency confirms that two very simplistic opposing visions were denied: firstly, the vision according to which global competitiveness is nothing more than a positive opportunity for growth and development. Secondly, the vision of globalisation as nothing else than a menace, a destructive force which strikes at the capability of the states and the EU to redistribute wealth and revenue. These two general visions present implications for the concept of European governance.

In the first vision, the only task of the governments would be to adjust by accelerating liberalisation. The EU should confine itself to the regulation of deregulation, what F. W. Scharpf, following Timbergen, has called 'negative integration'.[13] In the second vision, the priority granted to the defence of the existing social model translates itself into inward-looking policies of protection of sectors which are economically and socially fragile. These two cultures are well represented within the European societies as well as within the European Council. The Lisbon European Council, by launching a long-term strategy for 'the construction of the most-advanced knowledge-based society in the world', established that the globalisation process is not in itself destructive or beneficial. The consequences of the necessary opening to world competition will depend to a large extent on Europe's capacity to organise an adequate common political management of the modernisation process. It is through their strategic preferences, the mobilised resources, the coherent application of the policies adopted, that Europeans will succeed in taking advantage of globalisation. This all comes down to an original combination of negative and positive integration, which calls for reorganisation of the modes of

European construction. The question is not simply whether national systems of regulation can survive within the new framework of global competition, but what changes to the whole European system of governance, including the European State and other levels of authority, could better enable conception and application of an adequate strategic response. The expression 'governance' is used in this article in the sense given to it by J. Rosenau; the concept of governance is more comprehensive and encompassing than the concept of government. It includes both the formal and institutionalised procedures and the informal and non-institutionalised ones.[14] The concept of governance enables us to refer to more actors and levels of organisation of authority than central governments and their negotiations.[15]

Nevertheless, our concept of governance is evidently not limited to the non-state private dimension of authority (for instance contracts and partnerships, in which public authorities are solely perceived as 'enforcement to bargain'),[16] but takes directly into account the formal dimensions of the power of regulation of the state and of the sub-national levels in the framework of the 'organisation of collective action',[17] in the sense of 'provision of collective goods',[18] as well as the institutional and intergovernmental bodies of the EU.

These two dimensions of the current European governance are tied together in the framework of the system of European 'multilevel governance'. The scientific literature which shares this interpretation of European governance stresses in particular the following features: multiplication of levels and actors taking part in the decision-making process and in the implementation of the decisions; interaction between the sub-national, national, supranational and transnational levels of authority; continuous negotiation between interests at several levels, including public and private actors; the centrifugal, complex and overlapping character of the system and the absence of hierarchy in the organisation of authority.[19] To sum up, on the one hand, we have very intensive interaction between the European level and the national and sub-national levels, a true institutional interlinking, both in the sense of Europeanisation of national structures (top–down) and of the impact of national and infranational dimensions on the EU (bottom–up). On the other hand, the idyllic images of the multilevel governance are questioned: there is the absence of a true political community, or of a 'community of destiny', which makes this configuration, this polity, something 'intrinsically contingent',[20] still in transition towards its definite form. However unthinkable it may be to turn this multilevel governance into a European supranational State, it is not ruled out that internal pressures, external challenges and the will of actors may converge towards increased and new forms of solidarity and integration.

How does the 'Lisbon strategy' fit into the existing framework of European multilevel governance? The aim is to contrast the fragmented side of the European multilevel governance, according to a more coherent strategy and along three main guidelines.

Firstly, the strategic authority at the central level of the EU needs to be reinforced and the political leadership of the Union in this field to be restored. This leadership has to be strong enough to conciliate consensus between the fifteen, the Commission and the European Parliament around a long-term strategy (planned for a decade). The European Council (according to art. 4 TEU), in collaboration with the Council and the Commission, potentially represents this central authority.

Secondly, the reinforcement of the economic government of the Union was and is at stake, more particularly through the reduction in the above-mentioned asymmetry between the monetary and the economic union: this is only possible through strengthening the Broad Economic Policy Guidelines (BEPG) provided for by art. 99 TEC.

Thirdly, a method of governance had to be created which would include the Commission as well as the member states, and a multiplicity of institutional, social and economic actors. Only a new method, with a process which is both decentralised and centripetal, would make it possible to move forward in the process of convergence between European societies, despite the open opposition of euro-sceptical public opinions and of governments to new transfers of competencies.[21]

4.   THE LEGAL BASIS OF NEW METHODS AND THE MAIN FEATURES OF THE 'OPEN METHOD OF COORDINATION'

The new 'open method of coordination' (OMC) is at the heart of the 'Lisbon strategy' and its follow-up.[22] Despite the variations between different policy fields, this method has four components: 1. the establishment by the Council of common European Guidelines; 2. a reciprocal learning process, which includes benchmarking, peer review, diffusion of best practices and common indicators; 3. Given the quoted Guidelines and the learning process, national plans are drawn up by each government; 4. on a regular basis, the Council carries out an evaluation of the results, which can lead to recommendations. The Commission feeds the whole process with its contributions, both in the top–down and in the bottom–up dimensions of the strategy.

The precedents of the OMC are very diverse: first, the multilateral surveillance of the national economic policies, provided for in 1992 by the

Treaty of Maastricht (art. 99 TEC) in relation to the BEPG. Secondly, the approach launched by the Essen European Council (in 1994) in the field of employment, subsequently taken up by the Treaty of Amsterdam (1997, Employment Chapter, Title VII, art. 125–30 TEC) and precised by the Luxembourg European Council (1997), hence the name of 'Luxembourg process' given to the European Employment Strategy.

Indeed, the innovation of Lisbon is the result of the maturing process over several years. Even if the multilateral surveillance has evolved in its forms, there is a logic and an institutional spill-over effect in this development, including variations depending on the issues to which it is applied. The proposals to apply the same methods used in Monetary Union to the social and employment policies have been rejected many times by the European Council. The progress achieved by the Amsterdam Treaty employment chapter has to a certain extent pointed out a possible new direction for social governance. Article 128 again takes up art. 99 TEC, on the BEPG, but it is softer and much less binding. Given that the express goal of art. 99 is 'sustained convergence of the economic performances', the Council monitors the consistency of national economic policies with the BEPG and 'regularly carries out an overall assessment', 'on the basis of reports submitted by the Commission'. Its second specificity is the mechanism foreseen where it is established that 'the economic policies of a member state are not consistent with' the BEPG, and in particular, where 'they risk jeopardising the proper functioning of EMU'.[23] In that case, the process is gradually heading for procedures closer to the binding and centralised mechanism inherent to monetary Union. The Council can 'make the necessary recommendations', acting by qualified majority on a recommendation from the Commission. Following the same procedure, the Council can also 'decide to make its recommendations public'. Art. 99 confers the responsibility of reporting to the EP on the President of the Council and on the Commission, in particular in case recommendations are made public. Art. 104 C TEC foresees a possibility of 'escalation', ranging from the recommendation, to the public recommendation, 'measures' and also real sanctions (paragraph 11) applied to the member state responsible for an 'excessive deficit'. The 'Stability Pact', approved by the Amsterdam European Council (1997), following the request of the German Minister Waigel, has reinforced the binding character of the control of these budgetary policies.

However, in the Employment Chapter of the Treaty of Amsterdam, the mentioned mechanism of Maastricht providing for multilateral surveillance (art. 99) was inclined in the opposite direction, that of increased respect for national diversity. The 'Guidelines' adopted by the Council acting by qualified majority have a similar position as the BEPG of art. 99

(which they refer to explicitly), but they are based on the annual conclusions of the European Council. In comparison to art. 99, in general art. 128 confers greater importance on the member states with regard to the annual reports on the implementation of the national employment policies (the ones which will become, in practice, the National Employment Plans). Another difference is that whereas the European Council does not intervene in the control of the follow-up of the BEPG, left to the Council, the European Council receives the annual report from the Council and the Commission with regard to employment policies (examined yearly by the Council which, acting on a proposal from the Commission, can address recommendations to the member states). Art. 128 refers to the consultation of the EP, the Economic and Social Committee, the Committee of the Regions and the Employment Committee. Finally, art. 129 makes reference to 'best practices', comparative analysis, innovative approaches, evaluating experiences and pilot projects.

The Social Chapter is far more complex since the social objectives stated in art. 136 are placed in a framework conditioned, on the one hand, by the 'diverse form of national practices' and, on the other, by the 'need to maintain the competitiveness of the Community economy'. The possibility of having directives and qualified majority voting in the Council is foreseen in certain fields (gender equality, protection of workers' health, working conditions, information and consultation of workers), but with regard to the major dossiers of social protection (social security, etc.) the methods envisaged are cooperation and coordination: according to art. 137, the Council 'may adopt measures designed to encourage cooperation between member states through initiatives aimed at improving knowledge, developing exchanges of information and best practices, promoting innovative approaches and evaluating experiences in order to combat social exclusion'. However, according to art. 140, 'with a view to achieving the objectives of art. 136, the Commission shall encourage cooperation between the Member States and facilitate coordination of their action in all social policy fields under this chapter, particularly in matters relating to: employment, labour law and working conditions, social security, prevention of occupational accidents and diseases, occupational hygiene, the right of association and collective bargaining between employers and workers'. The ways by which the Commission 'facilitates coordination' between the member states are specified: giving opinions, arranging studies and consultations. Finally, as anticipated, arts 138–139 specify the forms of 'social dialogue', including the possibility of applying the very innovative procedure of 'contractual relations, including agreements', giving way to 'a Council decision on a proposal from the Commission' (art. 139, paragraph 2), should the partners so desire.

Thus, we are in the presence of a large multiplicity of quite varied regulation methods. Nonetheless, the common features are not negligible:

- The concept of 'coordination' as a method to achieve the EU's strategic objectives is used both in art. 99 (BEPG), art. 125 (employment) and art. 140 (social policy). However, it is not used in art. 104 and it is interlinked with the concept of 'cooperation' in the articles relating to social policy.
- The important role assigned to the Commission seems to go together with resorting to the concept of 'coordination' rather than to 'cooperation', a softer concept which is clearly more intergovernmental. The Commission plays multiple roles: (a) expertise and consultancy, but also more important roles, (b) preparation and periodical examination of national policies by means of reports (concerning art. 128, the report is a joint report by the Council and the Commission); (c) proposal of recommendations made to the Council and, when necessary, to the member states.
- The role of the Council, acting by qualified majority, both on the project and the recommendations (while acting by unanimity in the more sensitive fields of social security).
- The development of a Community culture of evaluation, exchange of best practices and so on is referred to both in the employment chapter and in the social one, within the framework of a kind of reciprocal and mutual learning process between the member states and encouraged by the Commission.

Even though the 'Lisbon Strategy' does not have a specific legal background covering all the fields concerned, it can benefit from a considerable legal acquis which justifies the step forward proposed by the Special European Council of March 2000. The Lisbon European Council has redrawn this complex legal background, enriched by the practical experience of the European Employment Strategy (Luxembourg process) and has formulated the new open method of coordination (OMC) putting it at the heart of the project of technological, economic and social modernisation for the next decade.[24] The new method was thus made more precise and applied (with adaptations as for its intensity) to other fundamental policy fields, traditionally under the competence of national and sub-national authorities: education, structural reform and internal market, technological innovation and knowledge-based society, research and social protection.[25]

We have already explained the meaning of the expression 'coordination' in the policy fields, which are, on the one hand, strongly affected by national diversities and, on the other, so important for the success of the EU's strategic common action, as the cooperation method alone could not produce the expected results.

Concerning the expression 'open', it implies several meanings, depending on the accents and interpretations:

1. Open in the sense of respect for the personalities, competencies and practices of the member states. The majority of member states call for a soft methodology because they are convinced that the great majority of the policy fields approached by the 'Lisbon Strategy' must remain essentially under the competence of national authorities and be submitted to unanimity in the Council. Hence, it is logical that the states be given guarantees when they are asked to accept the perspective of an increased convergence in these fields under their competence. The key argument: that benchmarking and the whole strategy respect the difference between (common) indicators and targets (different from one member state to another) is essential. Second, that the process of gradual approximation to the indicators be taken into account and not the absolute result.[26] The OMC does not foresee public recommendations to the member state whose policies are not consistent with the European guidelines, in contrast with art. 99, paragraph 4.

2. Open, in the sense of the participation of a variety of actors of the civil society: national, sub-national, transnational, supranational.

3. Open also, according to more integrationist points of view, because it might act as a *passerelle* towards other integration methods, in particular towards the Community method or social dialogue when deemed necessary. Following the same logic, this method could create the conditions to demand a subsidiary Community action, aiming to complete the action of the member states. Legal supports and precedents are not completely missing. Among others: 'minimum requirements' provided for in art. 137 paragraph 2; 'incentive measures' in the field of education, art. 149, paragraph 4; art. 125, according to which 'the Community shall contribute to a high level of employment by encouraging cooperation between Member States and by supporting, and, if necessary, complementing their action'. It would therefore be legally possible to fully respect the competencies of member states and at the same time go beyond cooperation, towards coordination and, if necessary, Community action. This last argument is, of course, very delicate because the OMC would clearly become not only a way to convergence, but also to deepening and integration.

The feasibility of the search for new methods of European regulation, intermediate ways between the Community and intergovernmental methods, is at stake here. As in the case of the European Employment Strategy and the Luxembourg process, among others, the EU is faced with an experimental method of social regulation, besides the European social dialogue (arts 138–139) and the legislative way (translated into directives).[27] A legitimate question that arises at the time of extending the new method to other and important fields is that of knowing if they are alternatives to the traditional methods or rather complementary methods. The answer remains open at this stage:

- On the one hand, the new methods are developed as complementary, acting where the Community method cannot be applied and the traditional intergovernmental method is not enough to achieve the convergence desired. It is therefore erroneous to interpret the OMC as a return of intergovernmentalism to the detriment of the Community method, since the method concerns policy areas which are not the exclusive competence of the EU. Moreover, we have emphasised the progressive interpretation of the expression 'open', and also the possible *passerelles* of the OMC to the Community method, if needed and when legally possible. The new method is not to be applied at the cost of European social dialogue either: on the contrary, the social partners have a role to play at each policy level (for instance the SEE) and at the global level (the annual *fora* of social partners launched in the year 2000 and continued in 2001). However, since the strategy has a global character (technological, economic and social modernisation), the EU cannot count exclusively on the methods provided for in the social chapter, the one applied to the BEPG, or the one of European Employment Strategy. To conclude: indeed, the open method of coordination has a specific character different from that of the directive, from the social dialogue or from other ways of regulation. However, it completes them in those cases where they are not applicable.
- On the other hand, the importance that the OMC and the new methods are gaining can have institutional spillover effects unfavourable to traditional methods. The EU could waive directives, even if they are possible from a legal point of view; or the OMC success could make the way for some issues from unanimity to qualified majority voting in the Council harder to achieve. However, such a development, if it were effectively to take place, would suggest that we are in the presence of a new integration regime in the making.

Apart from the words coordination and open, the expression 'global' should also be added to qualify the open method of coordination. Indeed, we should emphasise the importance of the express possibility of an extremely large application of the new methods to practically all the fields tackled by the 'Lisbon strategy'. These could be qualified as fields of 'common concern' (following the formulations of art. 126 and art. 99). It is precisely this 'global' character of the 'Lisbon strategy' that, to a certain extent, justifies the reference to the European Council as the political body that can ensure coherence and harmony between the different policies and between specialised Councils. The European Council, of course in collaboration with the Commission, plays a different role in the various policies: the European Council acts as a political guide in the case of the BEPG (submitted to the Conclusions of the European Council; see art. 99), an even more important role in the Employment Strategy, although marginal in the Social Chapter: as synthesis body and political leader, its role is crucial in the original government mode foreseen by the 'Lisbon strategy'.

5. GOVERNANCE WITHOUT GOVERNMENT? THE ISSUE OF STRATEGIC LEADERSHIP IN THE INTEGRATION PROCESS

The new methods of governance imply a new model of government. The majority of international scientific publications have come to the conclusion that the EC and the EU are unique cases of political systems, radically innovative in comparison to the classical institutional form of the nation-state. The logical implication of this is that the government of the EU – the same remark applies to its representation, democratic control, citizenship, and so on – cannot be analogous to the state models, or even to federal states.[28] The integration dynamics has increasingly stressed this *sui generis* feature of the European political system. This original character does not lie only in supranationality, but in the original combination of supranational, intergovernmental and mixed procedures and institutions. The 'constitutional' run-ups towards supranationality have regularly been accompanied by greater complexity and also by development of inter-governmental procedures. The main question in this contribution is to identify the forms the European government in gestation is taking, in other words, what kind of rebalancing the 'Lisbon Strategy' implies in this new power game between the different levels of authority and government in the Union.

The institutional innovation underway in the government of the Union is a double one: firstly, the global character of the OMC can only strengthen the centripetal demand for coherent coordination, capable of counter-balancing the articulation and decentralisation of governance practices arising from the 'Lisbon Strategy'. To a certain extent, the European Council, in close collaboration with the Commission, also applies the open method of coordination, not only through the establishment of synthesis indicators at the central level, but also through the institution of the Spring European Council, which shall take place once a year and which will be responsible for monitoring the state of progress of the strategy. On this basis, the European Council establishes an increased interaction (bottom–up and top–down) with the Commission, the social partners, the specialised Councils, the Committees, the EP, the ECB, and so on. To sum up, the innovative idea of 'periodic monitoring, evaluation and peer review' is the key of the central and decentralised government of the 'Lisbon Strategy', which distances it from the wishful thinking. Indeed, to wager everything on decentralised governance alone, without improving the organisation of the central and formal government of the strategy would be inefficient. The approach is decentralised, but the Commission, the Council and the European Council will ensure the coordination and control of the follow-up.

Secondly, the European Council, as the super network of prime ministers,[29] takes up again its strategic role of guidance by situating it in the framework of the essential goals of economic and social modernisation. The question of political leadership is open in the EU since the loss of strength of the classical federal/functionalist model of integration and because of the decline of the Franco-German engine. Several hypotheses have arisen both from the scientific debate and from political practices. The bet placed on the 'Lisbon strategy' is clear and many misunderstandings have gradually dispersed. The challenge to be able to catch up with the US and Japan in areas concerning the knowledge society, as well as to reconcile the economic and social logics in new forms of economic competitiveness and social justice, needed a political and strategic choice at the highest level of power. The most realistic balance between possible greater convergence and respect for national and regional specificities, the needs of the various states and sectors, could not seriously be conceived without an innovative attempt to reorganise the government of the Union: only the synergies between the European Council, the specialised Councils and the Commission can promote and strengthen the role of central political leadership and enable an inner redistribution of tasks.

The Lisbon and Feira Conclusions have insisted on the enhancement of 'the synergies within the central government' (European Council, Council and Commission) as a prerequisite for success. Paragraph 36 of the Lisbon Conclusions shows clear awareness of the difficulty of implementing the strategy in a coherent way: 'we should consequently be organised both upstream and downstream from that meeting' (the Spring European

Council). Of course, only the practical experience of the activities of the Commission and the successive Presidencies, from one Spring Special European Council to another can reveal if the hypotheses launched in the Conclusions are actually becoming reality.

The Commission, called to play an essential role, is the main ally of the European Council in the strengthening of the central government. The Commission is expected to take on multiple functions: it presents the priorities which will be at the heart of the European Guidelines to each Council and to the European Council; it organises the exchange of best practices and the peer review; it ensures the drawing up of the annual synthesis report on the eve of the Spring European Council in collaboration with the Council; it monitors the implementation and follow-up. Moreover, already since Maastricht the Commission is responsible for presenting to the Council the report at the basis of the BEPG, an essential element included in the new global strategy. Finally, since Amsterdam and Luxenburg (1997), the Commission draws the recommendations relating to the European Employment Strategy. The ECOFIN Council confirms its primacy as far as the BEPG are concerned, but the other Council formations are also called upon to contribute in an active way. In short, without a close collaboration between the Council, the European Council and a strengthened Commission, the 'Lisbon strategy' could certainly not move forward. Only the Commission possesses the technical and organisational means to ensure the continuity of the programme approved, and also coherence between Presidencies, which has proved to be a very delicate task, especially in recent years. It is nevertheless called upon to provide an enhanced internal coordination between its Directorates General and a 'political refocusing'[30] on the essential points of the strategy.

The Council is submitted to changes in its way of functioning. The General Affairs Council (GAC) could perhaps have aspired to an increased political role. But it is increasingly overburdened by the new tasks of the EU – as far as foreign and security policy is concerned – and cannot effectively take on that function. The reform of the GAC, wanted by many experts (breaking the GAC into two Councils: CFSP Council and Council for the coordination of European policies, consisting of deputy Prime Ministers), is not moving forward and would need fifteen convergent, analogous and simultaneous reform laws; in short, the harmonisation of the structures of the fifteen national governments, which does not seem easy to accomplish. Nowadays, on the one hand, the increasing weight of the stakes arising from the common foreign and security policy, and, on the other, the heavy work to coordinate incongruous policies prevent the GAC taking up this strategic role.[31] Without a reform of the Council, the practical relationship between the European Council, the specialised Councils and the COREPER encounters serious difficulties. The 'Lisbon Strategy' has opened the way to a possible development, which implies new relations of the European Council with the two most important specialised formations of the Council: the General Affairs Council, traditionally the most important one, and the ECOFIN which has gradually come first, since Maastricht, between 1992 and 2000.[32]

The valorisation of the European Council is worthy of comment. The double nature of the European Council – both of coordination and of general political and strategic guidance of the Union – is included in the Treaties: the European Council is a double body: on the one hand, it is the Council formation meeting at the highest level, that of Heads of State and Government; on the other, it is the body which covers the whole institutional system of the EU and the EC. This second aspect has been particularly important since the very beginning, according to a historical quotation, paradoxically coming from the inventor of the European Commission, J. Monnet at the founding of the European Council in 1974: 'the Heads of State and Government have decided to constitute themselves into a European provisional government'.[33]

Thus, in Lisbon, the European Council relaunched the primacy of the political body *per se*. The objectives set largely justified such an approach: they have a political nature and are essential to a civilian power such as the EU (contrary to the US, a politico-military power), which implies that the success of the modernisation of the social and economic model is a grand design, aiming at Europe's technological, economic and social catch-up *vis-à-vis* the US, but also, and, logically, the core of the development of its internal and international identity.

For some time now, the European Council has been working for a reinforcement not only of its role as supreme body of the Council, but also of its strategic leadership. The Helsinki European Council had already established that the BEPG should be under the coordination of the ECOFIN Council, but also 'under the political guidance of the European Council' (Helsinki Conclusions, December 1999). This implies restructuring the hierarchical relationship between the European Council, on the one hand, and the specialised formations of the Council, on the other, in particular the two which traditionally came first. Nevertheless, two questions arise. Firstly; would this institutional shift be worth a revision of art. 4 TEU?[34] Secondly, a theoretical question, what is the relevance of the primacy of politics at the start of the twenty-first century. The Presidency Conclusions from Lisbon assign a double function to the European Council in order to be coherent on its double nature: the European Council launches the strategy for socioeconomic modernisation and therefore mandates the specialised Councils and the Commission. Secondly, once a year, at its Spring meeting, the European Council is committed to coordinating the work of the Council and of the Commission, to monitoring it, to ensuring the follow-up and, if necessary, to relaunching the project.[35]

To sum up, the new link to establish between governance and government corresponds to a new model of organisation of the European powers: a reinforcement of the governmental function following a model of dynamic and oriented subsidiarity. The governance is decentralised, but the European Council, the Council and the Commission should constitute, each with its own role, the dynamic factors of guidance at the central level.

6.   TWO-LEVEL GAME OR MULTILEVEL GOVERNANCE? THE TRANSFORMATIONS OF THE EUROPEAN STATE IN THE PROCESS OF INTEGRATION

The Lisbon European Council does not propose a legal solution to the essentially political question of the sharing of competencies between the States, the Union and the regions. However, the concept of subsidiarity is above all a political one and does not as such result in legal conclusions in the framework of a relative lack of determination of Community competencies, where several competencies are mixed or shared.[36] The European Council is above all a political body and decides on a political dynamic, a 'centripetal orientation of subsidiarity'. The 'Lisbon strategy' and the OMC asked for a change in the role of States in the European integration, as well as within the member States themselves.

Indeed, we are witnessing a double transformation of the role of the States. Firstly, the States are increasingly networked (at the level of the European Council and of the Council formations) in the new political fields. Secondly, national governments are requested to carry out an internal reorganisation, in the sense of a greater strategic coordination between the various Ministries concerned and around the prime ministers. The OMC strengthens the State in terms of the need for internal coordination. Would its impact lead the EU to a radical rationalisation of the multilevel governance and to a return to the classic intergovernmental method of two-level governance?[37]

This would be too simplistic. National governments accept the need to engage themselves in a process which implies the establishment of European Guidelines, common indicators, a monitoring, and also, in certain cases, embarrassing recommendations from the Commission. Of course, they have a clear double interest: to improve national economic, social and technological performances and the efficiency of public policies, and above all to count on the external constraint (at least in certain countries) to help them gather a larger consensus around the most unpopular internal reforms. Governments can 'instrumentalise' the EU in the national public space, which has become a current practice in the EU. The 'Lisbon strategy' and the OMC result in a new international regime in which the limits and guarantees are widely decided by the States. This regime facilitates 'collective action by encouraging information flows, provides arenas for bargaining and establishes methods for monitoring and enforcing policies and contracts'. Even at the highest level, the European Council, composed of prime ministers (and their sherpas' network), constitutes a new engine of the European integration. Is it nothing but a mere new set of European intergovernmental regimes?[38]

Our answer is negative. By agreeing to take part in the new process, States are also submitted to deep transformations in their preferences, perceptions and respective interests, in the framework of a reciprocal learning process, leading to greater convergence. Moreover, they take risks by agreeing to be judged in this process, which includes supranational authorities and transnational actors and which deals with issues seldom dealt with at European level in the past. In cases where they are foreseen, recommendations from the Commission and the Council could increase the weight of criticism to governments from internal public opinion, economic and social actors, the parliamentary opposition. Last but not least, the result of this process will most likely be a greater convergence between European economic and social systems, and, what seems to us even more important, increased institutional interlinking between national, sub-national and Community authorities.

What is actually happening is a new development of the multilevel governance. It is becoming even more complex, with all that it implies in terms of mutual socialisation between national cultures and societies, but also in terms of increased complexity of the institutional and informal process of governance.[39] There is an increasing number of actors taking part in the multiple processes of coordination and submitting themselves to continuous interaction that can lead to increased European interlinking between the States and their political approaches. Governmental actors are to some extent obliged to work increasingly with each other within common institutions and to go beyond perceptions and actions which are strongly marked by national history. The evolving multilevel governance is indeed a real challenge: the Commission as a central actor could simply increase its tasks quantitatively, whereas the central level should undertake fewer activities and concentrate on the essential orientations. If the centre widens its distances from the national levels, failure is inevitable. There are numerous wrong ways: the process can be paralysed by the increase in fragmented complexity and the bureaucratic drift of the repetitive exercise. At best, the national public authorities are potentially implied in a mutual learning process at the European level with the goal to set common criteria, to improve the quality of the indicators, to spread best practices.

To conclude, there is an element of truth in both theoretical approaches mentioned: the new role of the European Council and of the Council can strengthen the intergovernmental networks as well as at the national level, as a facilitators of convergence, the coordination (around prime ministers) of the policy areas concerned. But governments are not the sole actors and States, who are themselves transformed by the new open method of coordination, must reorganise public powers. The question is: will the central strategic orientation and the centripetal tendencies be likely to control the most fragmented aspects of the European multilevel governance and to reinforce a new coherence of public action?

The numerous scientific works on the 'communicative and supervision State' provide excellent intellectual support for the hypothesis that, beyond functionalist and intergovernmentalist approaches, a deep transformation of the nature of public power itself is underway in Western Europe.

We have witnessed during the 1970s the crisis of political planning and the emergence of the theories of weak government and, in this context, the emergence of medievalist and hyperglobalising theories of the end of the State. Following the same intellectual dynamics, the Single European Act was interpreted as the beginning of the end of the European State, empty of substance, because of both internal and external processes: indeed, common policies have somehow gradually come to touch numerous fields of State competencies, border control, police, citizenship and immigration, currency, a part of taxation, government of the economy, industrial policy, representation and legitimacy, foreign and security policy. The OMC enables the last strongholds to be challenged: employment, education, research, welfare standards and knowledge society. Classical State sovereignty has completely fallen apart in the globalised world and in particular in the context of European integration.

However, the question of the future and the changes to the European State remain open, and the 'Lisbon strategy' could be placed within this debate. Indeed, we are facing a plurality of national experiences, and some States adapt to the implications of increased interdependence and to the new European and global economy context by developing a new concept of public policy. How? Either in the name of the comparative advantages of nations, as competitor States, or in the perspective of European cooperation, coordination and integration. The 'Lisbon strategy' has relaunched this second option by emphasising politics, the coordination of public policies and positive integration along with negative integration. It breaks with the Saint-Simonian dream of 'non-political progress': technology and market logic would lead to the weakening of the State and to the transfer of loyalties to the post-national community. A disinterested elite would act as a driving force. The 'Lisbon strategy' breaks with an anti-political – and not only anti-State – idea of European integration.

New concepts of public action within a knowledge society, for instance that of supervision State, become current.[40] Public power is seen as a multiplier of communication via the coordination of administrations. It organises the decentralisation of execution, but, at the same time, it highlights the function of strategic synthesis. In this last case, we witness an interpenetration of governments and administrations across Western Europe, that goes clearly beyond what is described by the intergovernmental approach. Maybe without idealism the States are adapting to the implications of informal economic and social integration, to the imperatives of technological innovation and to budgetary constraints, and are discovering a new dimension of public service.

7. OUTPUT LEGITIMACY AND EUROPEAN PUBLIC SPHERE

The 'Lisbon strategy' is based on the conviction that the efficiency of economic and social policies is a worthy criterion of legitimacy for the EU (output legitimacy).[41] This opinion is shared by most international scientific literature. According to J. Weiler the 'substantial' legitimacy of European integration reached its peak when the procedures of formal legitimacy were the least developed, that is during the 1960s and 1970s. Public opinion identified the European construction with progress, economic prosperity and full employment, because the integration process was complementary to the construction of the national welfare state systems based on Keynesian policies and national growth compromises. Mass unemployment in the 1990s would be one of the causes of euroscepticism because the EU would be considered not only as inefficient against unemployment, but also as one of the causes of the fall in the employment rate.

Can we expect the positive dynamics of the 1960s and 1970s to reproduce themselves in a post-Keynesian period?[42] Definitely not with the same balance between national diversities and European convergence as in the past, neither with the same kind of 'complementary distance' between the national and European dimensions of the economic and social modernisation policies. Within the new modernisation strategy, the Commission in collaboration with the Council is called upon to increase convergence by coordinating national policies. Should this fail, there would be an increased temptation to multiply the independent agencies as a kind of alternative to the Commission. The efficiency of central performances will be key to legit- imising a renewed European socioeconomic model.

However, numerous problems may arise. Indeed, the multiplication of the Council's and the Commission's tasks in the name of the efficiency of policies (there are around fifty additional tasks consecutive to the Lisbon European Council) will not be without consequences as far as the balance between effectiveness and legitimacy is concerned. The legitimacy crisis, in the sense of lacking social consensus, runs the risk of getting worse.

Secondly, the benefits of the OMC risk being monopolised by the States, which blames the EU for the unpopular constraints and takes credit for the positive consequences of the successes, especially in terms of the domestic legitimacy of governments.

Thirdly, the European political culture of participatory democracy and democratic scrutiny questions the new methods demanding innovative forms of input legitimacy. It would certainly be wrong to expect that the legitimacy deficit could be absorbed simply through improvement in the procedures of conventional political legitimacy. Indeed during the 1990s public opinion became more eurosceptical and attached to the symbols of national and local sovereignty, despite both the increase of the co-decision powers of the EP and the establishment of a European citizenship. However, the new role of the European Council and the Council raises crit- icisms because of the marginal role of the EP and the increased bureau- cracy of the repetitive annual exercise. They need an answer to be given.

As far as the European Council is concerned, its national legitimacy is very strong because it is based on democratic national elections of govern- ments and prime ministers. National Parliaments have the right to ensure that national plans based on European Guidelines are applied, and increase their control over their respective governments' European policy in these crucial areas. However, if the legitimacy of the 'Lisbon strategy' is only national, it does not correspond to the supranational character of the process. Since it is a European strategy, legitimacy should also be suprana- tional. The first question would therefore be to know through which enhanced role of the EP supranational legitimacy could be reinforced. The practical solution found by the Portuguese Presidency is very interesting: for the first time, the Report of the Presidency of the Council to the European Parliament took place not only after, but also before, the European Council and the EP approved a favourable Resolution.[43] However, this formula has no legal basis and nothing could force the fol- lowing Presidencies to repeat the successful but risky performance of Mr Guterres and the Presidency before the EP. The question could be raised at the level of the reform of art. 4 TEU, to establish an annual plenary session of the EP, devoted to the socioeconomic strategy of the EU, before each Spring European Council.

The OMC is a learning process and it will take some time before social acceptance of an increased convergence is acquired. We could witness a risk of the breakthrough – multiplied by the number of policies concerned – of the critical statements against 'neocorporatism and lacking transparency' which emerged from the EP against the European Employment Strategy in October–November 1999 and in 2000. This criticism is backed up by some member states. Indeed, the socially and economically organised actors see their participation increase (within Committees, *Fora*, etc.) even though they criticise the absence of a central place where concertation on all the socioeconomic processes underway would take place. The proposal to organise a Forum of social partners before each Spring European Council was only accepted case by case. If the actors of civil society are not concerned, consulted, implied as partners and negotiators, one of the characteristics of the openness of the new method will be denied. But the mere participation of socially organised actors is far from a guarantee of as great a legitimacy as in the past, because of the representation crisis of the social organisations and the increasingly fragmented lobbying of interest representation.[44] A response to these worries expressed in the name of European citizenship (which would increase the EP's role of democratic scrutiny in relation to that of co-decidor) would be to establish the rule of regular contacts between the Presidency of the Council and the specialised Committees of the EP, with regard to the progress of the different chapters of the 'Lisbon strategy' and the application of the OMC.[45]

In general, we would like to conclude with a *caveat*: even though the request for greater social and political participation in the process of the open method of coordination is very diffused, its practice could not be 'the panacea'. As such, it could also be a question of rhetoric and the paradox of participation could even lead to the paralysis of the process (because of the over-commitment of requests and of their incremental effect), in the absence of a correct balance between governance and government. New forms of interaction and efficient communication between the new methods of governance and civil society should be created, and the EP should strengthen its synthesis role.

In conclusion, it is clear that the failure or the paralysis of Lisbon's ten-year programme would have heavy consequences in terms of the EU's legitimacy in public opinion, which would also entail a decay of the European construction. However, the success of this strategy would reinforce the essential core of the democratic perspective at European level, that is the practice of the values of justice and efficiency; thus, the search would be encouraged for a third way between local and national narrow visions of democracy, on the one hand, and, on the other, the globalist utopias on cosmopolitan democracy.[46] However, the perspective would be that of a new system of democratic legitimacy and governance: multilevel (international, national, supranational, transnational), multifaced (territorial, functional, modern and post-modern) and with a multitude of actors (social, economic, political and cultural; institutional and extra-institutional), rather than that of a classical democratic normative model – federal/constitutional[47] or democrat/republican.[48] Whether the result of this complex interaction between the dynamics of governance and the dynamics of participation will be a social regime characterised by fragmented lobbying of the groups and the privatising of governance[49] or an innovative model of citizenship and democracy, both state and post-state, is a question that remains open. Of course, an innovative result will not only be the effect of decentralisation of governance in comparison to the traditional State regulation as the optimists see it;[50] it will also depend to a great extent on the success of the dynamics of horizontal and centripetal coordination, as well as on the new methods of government, that is, the orientation given to the multilevel governance.

A new focus for research has appeared: the question of the place of participatory democracy. It should be placed within the framework of the dialectics between the depolitisation of domestic cleavages as the inevitable effect of the open coordination strategy launched in Lisbon, on the one hand, and, on the other, the development of a transnational public space, including a real debate (between centre-right and centre-left interpretations of modernisation), as a relevant pillar of European transnational politics.[51]

8. THE 'LISBON STRATEGY' AND THE POLITICAL QUESTION OF THE FUTURE OF EUROPEAN INTEGRATION

The 'Lisbon strategy' revives the concept of the primacy of Politics against a mere administrative and technocratic concept and the practice of piecemeal governance by fragmented authorities. The current decline of Politics was – and is – about to make relevant political decisions impossible both at national and infranational levels. Relaunching political design and reform is a tremendous challenge facing huge obstacles, concrete antipolitical streams, new medievalist and hyperglobalising tendencies.

Secondly, the Lisbon decision is an affirmative and long-term European 'grand strategy', with the ambitious objective to update the European social model and its identity in relation to global competition. It offers an array of new modes of governance, adapted to the issue to be solved. Its potential is enormous in terms of increasing strategic and real convergence and of mobilising national budgets through the open coordination of national and regional policies.

However, the implementation of the 'Lisbon strategy' within the process of European integration is facing three main problems:

1. Although the Portuguese Presidency was proactive, long-term oriented while able to compromise to obtain an unanimous decision in Lisbon, the Presidency, particularly that of a relative small country, has little room to manoeuvre, which has consequences for its implementation.

   A structural limitation is provoked by the contradiction between the very large scope of the strategy and the very short time available to coherently and consistently implementing it. The continuity between Presidencies is assured by the European Commission, but the priorities of Member States and rotating Presidencies are frequently different for geopolitical, prestige, political, cultural and economic reasons. True, the Conclusions of the Nice European Council (December 2000) do often cite the 'Lisbon strategy' as a global framework for socioeconomic policies, include the new 'social Agenda' within the 'Lisbon strategy' and call for the Stockholm Spring 2001 European Council to succeed.[52] However, in terms of economic governance and government, the priorities of the French Presidency have been the strengthening of the 'Eurogroupe and the IGC', with controversial results. Indeed, the consistency of the Swedish, Belgian, and Spanish Presidencies will be crucial for the success of the 'Lisbon strategy'.

2. A 'grand strategy', to combine legitimacy and effectiveness, would require considerable media investment and communication with European public opinion, consistent with the much proclaimed awareness of the rules of an information society. The question is whether the European Commission will engage its extraordinary experience in campaigning for common European objectives, including 'public anchors' deadlines, indicators, media and social mobilisation as it did, for instance, with the famous and successful 'Europe 1992' campaign (1987–92).

3. The 'Lisbon strategy' and the Open Method of Coordination are part of broader European research on governance and government within the enlarging EU, which can hardly be increasingly divided into many independent parts. Of course there are different levels of the evolving European construction. But the institutional, legal and constitutional implications of the strategy and the OMC should be elaborated in order to provide an original contribution to the triple post-Nice agenda: enlargement, European Constitution and institutional

reform,[53] including 'enhanced cooperation', share of competencies and political perspectives of European integration.

The contribution already provided by the Portuguese Presidency entails very new answers to such highly controversial questions: 1. to foster European convergence from the current formal criteria of Monetary Union towards deeper societal and economic convergence; 2. to practice open coordination of national policies of all Member States rather than through hard cores and directorates; helping the States and public authorities in their transformation as modern supervisory bodies rather than fixing regional, national and European competencies.

NOTES

1. Ian Clark (1997), *Globalisation and Fragmentation*, Oxford: Oxford University Press, J.N. Rosenau (1995) 'Governance in the Twenty First Century', in *Global Governance*, 1, pp. 13–43 and C. Crouch and W. Streeck (ed.) (1997), *Political Economy of Modern Capitalism. Mapping Convergence and Diversity*, London: SAGE. See also R. Gilpin (1993), 'Welfare nazionale e norme economiche internazionali', in M. Ferrera (ed.) *Stato sociale e mercato,mondiale*, Torino: Fondazione G. Agnelli, and A. Prakash and J.A. Hart (ed.) (1999) *Globalisation and Governance*, London: Routledge.
2. James N. Rosenau (1999), 'A transformed observer in a transforming world', in *Studia diplomatica*, n.1–2, edited by C. Roosens, M. Telò and P. Vercauteren, pp. 5–15.
3. Ph.G. Cerny (1997), 'Paradoxes of the Competition State', in *European Journal of Political Research*, n.3.
4. See the *Introduction* to this volume by Maria João Rodrigues and particularly the paragraph, 'Europe at the cross road'.
5. Preamble of the Treaty on European Union, (TEU).
6. Preamble of the Treaty establishing the European Community, (TEC).
7. Ph.G. Cerny (1997), cit.
8. European Commission (1993), 'Growth, Competitiveness, Employment White Book', *EC Bulletin*, supplement n.6, 1993.
9. Recently A. Moravcsik, (1998) *The Choice for Europe*, Cornell, Ithaca.
10. The expression is from J.H. Weiler (1996), 'Comunità europea', in *Enciclopedia delle scienze sociali*, Roma. The 1st number, 2000, of the *Journal of European Integration* is entirely dedicated to the analysis of euroscepticism.
11. The European Commission has already expressed its openness to the 'new methods of governance' in its document of March 2000, which is, among others, at the source of the vast and thorough work of inventory and analysis of the transformations underway in the reality of the EU at the core of the White Book on Governance, (July 2001).
12. W. Streeck (1996), 'Neo-Voluntarism: a New European Social Policy Regime?', in W. Steeck and others (eds), *Governance in the EU*, London: Sage.
13. F.W. Scharpf (1997),*The Problem Solving Capacity of Multilevel Governance*, Florence: IUE, and G. Majone (1996), *Regulating Europe*, London: Routledge.
14. J. Rosenau & O. Czempiel (1992), *Governance without Government. Order and change in World Politics*, Cambridge: Cambridge University Press.
15. A. Prakash and J. A. Hart (1999) cit., pp. 310–17.
16. D. Lake (1999), 'Global Governance. A Relational Contracting Approach', in A. Prakash and J. A. Hart, cit., pp. 31–53.

17. See O. Young (1986) 'International Regimes: towards a New Theory of Institutions', in *World politics*, 39, pp. 104–22.
18. A. Prakash and J.A. Hart (1999), cit., pp. 1–24.
19. Ph. Schmitter, G. Marks, F. Scharpf, W. Streeck (1996), *Governance in the EU*, London: Sage, B. Kohler-Koch (1998) (ed.), *Regieren in Entgrenzten Räume*, Opladen: Westdeutscher Verlag.
20. Ph.C. Schmitter (1996), 'Imaging the Future of European Polity', in Ph.C. Schmitter and others, *Governance in the EU*, cit, p. 150.
21. See the chapter of this book, by M.J. Rodrigues, on this crucial point.
22. For the detailed description of the OMC, see the documentation annexed and particularly the Conclusions of the Lisbon European Council and the paper approved by the Feira European Council on the OMC; see also the introductory chapter by M.J. Rodrigues.
23. Concerning art. 99 (ex art. 103), it is obvious that the multilateral surveillance mechanism is too heavy (two readings by the Council are provided for, before and after the 'conclusion on the BEPG' adopted by the European Council). The first reading could simply be replaced by the report of the Commission to the European Council. Indeed, the sharing of the work between the European Council and the ECOFIN, and the role of the other specialised Councils ought to be precised. The European Council is evoked only once. It is nonetheless the body of impulsion and strategic decision (by unanimity), the political guide of the process, whereas the ECOFIN is a body where legal, formal (by qualified majority) decisions are taken. As for the role of the Commission, it is still too limited at the level of the implementation. It should also be allowed to ask the Member States for information.
24. *Presidency Conclusions* (Lisbon European Council), Ref. Council of the EU, SN 100/00; *Presidency Conclusions* (Santa Maria da Feira European Council), Ref. Council of the EU, SN 200/00; Presidency of the EU, *Note on the ongoing experience of the Open Method of Coordination*, Ref. Council of the EU, 9088/00, 14.06.00.
25. The evolution of the Council Education is particularly important because the practice of the new method has enabled the development of indicators and new programmes in the fields of lifelong learning and the reconceptualisation of old programmes, like the programmes to encourage the mobility of students (which can be strengthen thanks to national budgets. See Nice European Council Conclusions). The Council Research has also launched a process coherent with the 'Lisbon strategy'. Significant innovations are also underway concerning the Council Employment and Social Policy (see Council ESP of 18 October 2000, and secondly, the approval of the Social Policy Agenda in November). Concerning the potential of effective cooperation in the field of the fight against social exclusion and of social protection, see paper from the Belgian Minister for Employment, Frank Vandenbroucke, *Toward an European Social Policy: Turning Principles of Cooperation into Effective Cooperation*, London, 11.11.2000. See also the concluding remarks of the chapter provided here by Gøsta Esping-Andersen.
26. See chapter of the book by B.A. Lundvall and particularly his remarks on the political dimension of common indicators and of the process of setting them.
27. J. Goetschy (1997), in Telò/Magnette (ed.), *De Maastricht à Amsterdam*, Brussels: Complexe. According to J. Delors, 'the economic and social resolution adopted in Lisbon is on the right direction'; 'it is interesting to see that, in the economic and social field, we are searching for new methods of governance: the directive is not appropriate, simple concertation is not enough' (Hearing of J. Delors before the delegation for the EU of the French Sénat, Paris, April 5 2000, in Agence Europe, 28.4.2000, pp. 3–5).
28. M. Telò (1999) (ed.), *Démocratie et construction européenne*, Brussels: Editions de l'ULB.
29. And the network of Prime Ministers' Sherpas, who held four informal meetings under the Portuguese Presidency, in preparation for the two European Councils.
30. This expression comes from Report of the Commissariat Général au Plan, Group lead by J.-L. Quermonne (1999), *L'EU en quête d'institutions légitimes et efficaces*, Paris: La Documentation française, pp. 66–7. It is also one of the main guidelines of the Commission White Paper on European Governance (July 2001).

31. See the explanations given in the Hearing of Mr De Boissieu – Deputy Secretary General of the Council – before the Commission of constitutional affairs of the EP, March 2000.

32. See Fiona Hayes-Renshaw and Helen Wallace (1997), *The Council of Ministers*, New York: St Martin Press. See J.P. Jacquet and D. Simon, 'The constitutional and juridical role of the European Council', in *The European Council 1974–1986:evaluation and prospects*, J.M. Hoscheit and W. Wessels (eds) (1998), IEAP, Maastricht and J. Cloos, G. Reinesch, D. Vignes, J. Wyland (1993), *Le traité de Maastricht, genèse, analyse et commentaires*, Brussels: Bruylant.

33. This statement has to be placed in the framework of the European debate which preceded the transformation of the Summits into a European Council (1974). See J. Monnet (1976), *Mémoires*, Paris: Fayard, pp.591–2. I thank M.C. Franck for having drawn my attention to this interesting chapter of the *Mémoires*.

34. Concerning art. 4 (ex art. D), the practice of the Portuguese Presidency has constituted a dynamic compromise between two visions: on the one hand, the maximum of external primacy in relation to other European institutions was achieved at the Special European Council (significantly commented upon using the French words 'L'Etat c'est moi'. See A. Manzella, in *La Repubblica*, Roma, 25.3.2000, p.10); the European Council refused the previous drift towards some kind of Court of appeal of the Council and reversed the pyramid of the institutional practice taking up the role of guidance in the process. The European Council covers the three-pillar structure and all the activities of the EU (art. 3: 'the Union shall be served by a single institutional framework'). On the other hand, concerning the implementation of the 'Lisbon strategy' and its continuity in time, the Presidency and the European Council have developed an increased interaction with the EU/EC institutional system, particularly with the Commission.

The decision to establish a Spring European Council is capital for this purpose. It imposes itself as an economic government in relation to the strategy of economic and social modernisation.

The two logics are not really completely opposed. A reform of the Treaty could retain, in art. 4, the formula used in the Stuttgart Solemn Declaration of 1983 (refused in Maastricht): 'when the European Council acts in the Community matters, it does it as the Council, in the sense of the Treaties'. Or: 'when the European Council acts as economic government of the Union, it does it as the Council, in the sense of the Treaties'. Indeed, the decision for the beginning stage 3 of EMU constitutes the most important precedent concerning the role of economic government of the European Council: the quite unusual formulation used in arts 121 and 122 offers a third possible track: 'the Council meeting in the composition of Heads of State or Government'. This formulation was considered an institutional innovation, *ad hoc*, a *sui generis* compromise (D. Simon). But the drawback is that even if the acts adopted shall not be considered as Community acts (with all implicit consequences), this idea could, if applied *in extenso* to the role of economic government of the European Council, reduce its role of political and strategic boost. It is therefore better not to go that far, even in the case of an approximation between the European Council and the Council.

35. 'Le Conseil européen s'est autoproclammé gouvernement économique', these were J. Delors' comments. 'By reinforcing the political counter-balance of the ECB, the European Council has thus brought his own answer to the problem of leadership in the EU', has insisted the Swedish Prime minister, G. Person (*Agence Europe*, March 2000).

36. This is, in part, the interpretation put forward by the Report of the Commissariat au Plan (cit., pp.99–100) which states that 'the EU cannot be happy with the mechanical and ritual call upon the principle of subsidiarity. In the absence of a list of competencies the EU has to call on other methods'.

37. For this theoretical approach, see A. Moravscik (1998), cit.

38. S. Krasner (1983) (ed.), *International Regimes*, Cornell, Ithaca; Moravscik (1998) cit.

39. 'La politique européenne de l'emploi: réflexions sur les nouveautés de 1999 et leur impact pour la Belgique', in *Revue belge de sécurité sociale*, 2000.

40. H. Wilke (1997) *Supervision Staat*, Suhrkamp; S. Unseld (1993) (ed.) *Politik ohne Projekt?*, Suhrkamp (in particular J. Esser, *Die Suche nach dem Primat der Politik*); R.

Voigt (1998) (ed.) *Des Staates neuer Kleider*, Nomos; Kohler Koch (1998) (ed.), *Regieren in entgrentzen Räumen*, Opladen.

41. F. Scharpf (1998) cit.
42. See the contribution of R. Boyer published in this book.
43. European Parliament, *Resolution of the European Parliament on the Lisbon Special European Council*, Ref. B5-0236, 0239 and 0240/2000, March 2000.
44. G. Falkner (1998), *EU Social Policy in the 90s*, London: Routledge; H. Wallace and A. Young (1997), *Participation in the European Union*, Oxford: Oxford University Press; G. Feyertag and Ph. Pochet (eds) (2000), *Social Pacts in Europe: New Dynamics*, ETUI, Brussels.
45. It is obvious that the respective roles and competencies of the Commission, the ECB and the organised social actors also ask for a clearer definition. The EP has put forward a proposal already in its March 1999 Resolution, particularly concerning the procedure of the BEPG (art. 99, par. 5). The EP expresses more and more the feeling of being marginalised in a large part of European governance.
46. D. Held (1995), *Democracy and the Global Order. From the Modern State to Cosmopolitan Governance*, Stanford, California.
47. J. Habermas (1991), *Die post-nationale Konstellation*, Frankfurt am Main: Suhrkampf.
48. R. Bellamy (1999) 'Una Repubblica europea?', in M. Telò, 'Quale idea d'Europa per il XXI secolo?, *Europa, Europe*, n.5.
49. H. Wallace and A. Young (1997) cit.
50. O. Czempiel (1992), in J. Rosenau and O. Czempiel, *Governance without Government?*, cit., pp. 250–70.
51. The 'Lisbon strategy' has a timely perspective and calls for policies for the medium term, policies independent from the internal contingencies (electoral, etc.). It depoliticises in part the stakes (unemployment, social protection, fight against the digital divide, etc.) as internal electoral cleavages, even though it politicises the fight for employment and social modernisation as a European common stake. We have stressed the highly political character of the 'Lisbon strategy', both on what concerns its importance for the autonomous identity of the EU at world level and on its implications for the reorganisation of internal powers (hierarchy within the Council). Of course, it cannot just be considered as a centre-left strategy. Indeed, it is the result of a European Council, conditioned also by the political beliefs of its Presidency and its majority. Nonetheless, the approval and the implementation of the 'Lisbon strategy' are both new grounds of confrontation and collaboration between the centre-right and the centre-left, in a pluralist Europe. Governments of centre-right, like Mr Aznar's, or which do not represent the traditional continental social democracy, like Mr Blair's, have played a very important role. Inevitably, we witness a large variety of interpretations of the common strategy. The unions' vision (ETUC) of the OMC (of its fields and modes of application) is not the same as UNICE's. However, the extreme view points are not encouraged. It's a bit like what happens with the national social pacts. We have social pacts managed by the centre-left and social pacts managed by the centre-right. Both centre-left and centre-right contribute to it and the solutions are often a compromise between heterogene forces.
52. See *Bulletin Europe*, 12.12.2000, European Council, *Nice: Presidency Conclusions*, p. 10.
53. The practical shifts in governance made on the Presidency's initiative have not interacted at all with the 2000 intergovernmental Conference on the institutional reforms to be carried out in view of the enlargement. Nevertheless, the 'Lisbon strategy' does not exclusively concern the policies, but also the EU's polity. As in the past, alongside the official IGC, another 'practical IGC' took place, which produces factual changes parallel to the negotiations between the States concerning their power relations. These important shifts and institutional innovations presented however implications for the long-term process of Treaty reform. The 2000 IGC will indeed not be the last, a new one is already scheduled for 2004. The aspects of the Lisbon Strategy essential to the future of EMU and to the European socioeconomic model, presenting general implications for the EU's governance and government, ought to be included on the agenda of the next IGCs.

Annexes

Portuguese Presidency of the European Union

Document from the Presidency

Employment, economic reforms and social cohesion – towards a Europe based on innovation and knowledge   Lisbon, January 2000

1. COORDINATING EFFORTS TO ACHIEVE A NEW STRATEGIC GOAL

1.1. A new strategic goal

Europe, as it enters the new millennium, demands a new vision and a long-term strategy. While retaining all that is best in its traditions and values, Europe must develop as a civilisation which bases its economic and social prosperity on the advancement of knowledge, cultural diversity and cohesion and which plays an active role in promoting a more balanced, peaceful and harmonious world order.

A new strategic goal needs to be defined for the next ten years: to make the European Union the world's most dynamic and competitive area, based on innovation and knowledge, able to boost economic growth levels with more and better jobs and greater social cohesion.

1.2. An affirmative strategy

A new period is beginning in the process of European construction. The European Union is gaining substance economically, socially and politically. Initiatives in the area of common foreign and security policy and a common area of freedom, security and justice add to the great achievements of the single market and the single currency. The enlargement process too is now in full swing. These are major historical milestones in the affirmation of the European project.

Despite the economic recovery, serious social problems continue to exist, such as unemployment, social exclusion and the risks of future imbalance in social security systems – which are also the reflection of deeper-seated structural difficulties calling for bold reform. These difficulties are heightened by the unavoidable challenges posed by globalisation, technological change and an ageing population.

The economic and social strategy of the European Union must not be devised solely as a defensive response to these challenges but as an affirmative and creative response to the new opportunities which are emerging. This means redefining Europe's role in the world economy, building a new competitive platform, opening the way for new and better jobs and organising this movement with social cohesion.

It is essential that we regain the conditions of full employment geared to the needs of the emerging society, more open to the options of European women and men. This calls for the creation of a growth dynamic ensuring a sustained average annual rate of at least 3% for the whole of the European Union.

1.3. Towards a new paradigm

Macro-economic stability is fundamental to consolidate the euro and ensure sustainable growth. It is also essential now to foster a culture of dynamism and entrepreneurship and a culture of strengthened social cohesion. The current improved economic situation is the right time for the necessary reforms to be undertaken.

Throughout the world nations seek to progress while maintaining the difficult balance between openness, diversity and cohesion. In the context of globalisation companies are increasingly defining their strategies in world terms and capital movements are increasingly being controlled by the major financial centres according to the level of national and business competitiveness. This is increasingly dependent on the capacity to provide a swift and innovative response to the more individual needs of the market, and calls for the generation and dissemination of a vast store of knowledge, made possible by the ongoing revolution in the field of information and communications technology.

A new paradigm is emerging, encompassing both a technological revolution and a major change in the social exchange of knowledge affecting all institutions, from schools to businesses and from public services to the media. The transition to an innovation- and knowledge-based society and economy is under way.

Innovation and knowledge are increasingly becoming the decisive source of wealth and also the main source of difference between nations, businesses and people. Fresh opportunities for redefining European competitiveness and creating new jobs are thus arising, but also new risks of social exclusion.

1.4. Coordinating policies

Despite a number of undeniable successes, Europe is lagging behind in this transition to the innovation- and knowledge-based economy. This delay is apparent in the production and dissemination of much information technology but also in adaptation of social institutions and relations to the new potential opened up by such technology. While this failure to adapt to the new paradigm continues, there will be a shortfall in economic growth and an increased risk of unemployment and social exclusion.

We need to increase the pace of technological change but also of institutional reform and to learn best practices more quickly, but also to create new best practices. Innovation in political method is also necessary.

An economic and social strategy to renovate the basis for growth in Europe must combine macro-economic policies, economic reform and structural policies, active employment policies and the modernisation of social protection.

Institutional processes for the development of these policies, namely the Cologne process on macro-economic policies, the Cardiff process on structural policies and reforms and the Luxembourg process on employment policies, are now available to the European Union. Our intention is not, therefore, to launch a new Lisbon process.

However, we do see the Lisbon European Council as a particularly good opportunity to create the conditions for:

(a) articulating, simplifying and extending the existing processes through improved coordination, in order to achieve a new strategic goal
(b) adding new dimensions in key areas such as preparing for an innovation and knowledge-based economy, combating social exclusion and modernising social protection
(c) developing coordination methods for formulating, quantifying and monitoring policy objectives and instruments.

These methods may vary. For example: in the case of social protection, joint analysis, cooperation and the exchange of best practices; or, in the case of the information society policy, the definition of European guidelines, national plans and a benchmarking process with reference indicators permitting intra-European comparison and a comparison with other areas. This involves an open method of coordination coupling coherence with respect for national diversity. It also involves learning how to respond more quickly to structural change.

The development of these methods will need backing in its various stages from the European Commission, as an essential catalyst.

Europe must find its own way of constructing an innovation and knowledge-based society and economy. A rich scientific and cultural heritage and an immense capacity for generating new knowledge are available to it. The European way needs to open up opportunities for accessing knowledge, value cultural diversity at its true worth and use this transition in order better to forge a specific European identity and to identify citizens more closely with a European project which they themselves will define.

A way based on the gradual construction of a European public area, the exercise of European citizenship and the promotion of dialogue with the various actors, with the emphasis on social partners. The High Level Forum in June, bringing together representatives of Governments, the European Commission, the European Parliament, social partners, the Economic and Social Committee and the European Central Bank, will reflect this approach.

A major transformation is under way in Europe. It is for us Europeans, with our creativeness and political will, to endeavour to lend it shape and form.

2. TOWARDS AN INNOVATION AND KNOWLEDGE-BASED ECONOMY AND SOCIETY

Faced with the digital revolution Europe, like the United States, initially focused its response on information technologies, then on the information highways, and subsequently on the information society. Nowadays it is becoming clear that the problem is not only about information, but about knowledge and innovation, and not only about technological change, but also about economic and social change.

All societies are knowledge-based. What is new is that the information and communication technologies are changing the way in which knowledge is accumulated. More and more knowledge is being built into equipment, products and services. Knowledge is increasingly becoming the raw material of work. However, this newly emerging model is still giving rise to many dilemmas: how to develop the strategic segments of the new value chains taking shape worldwide? How to make room for cultural diversity in the cyberspace now being built? How to stimulate innovation, not only in processes, but also in products and services, in order to boost job creation? How to equip the workforce for much more rapid changes in occupational activities? How to cope with the new social inequalities?

Thus, given the wider implications of the new paradigm and the dilemmas created, the European strategy to be defined must:

(a) create a demand-driven dynamic stimulating innovation in products and services, meeting citizens' requirements and influencing technological choices on the supply side;
(b) create another competitive platform in infrastructure, hardware and software, available knowledge, entrepreneurial capacity and additional skilled jobs – which requires a major boost of intangible investment;
(c) mainstreaming the concern for social inclusion;
(d) play a pro-active role in organising cyberspace (on-going negotiations on e-commerce, register of Internet fields).

Moreover, the widespread development of scientific and technical skills must be recognised as a key factor of employment policy in Europe. The consolidation and updating of scientific and technical skills and the widespread acquisition of IT skills are central to the creation of skilled employment and the construction of a competitive economic and social base. In order to achieve these objectives, the importance of a scientific and technological culture for the entire population needs to be highlighted as does the need for far-reaching scientific and technological development supported by a strong and open European R&D policy.

In this wider framework the policy for an information and knowledge-based society cannot be dissociated from S&T policy, nor from the policy on education and training, and must also be linked to the policies aimed at supporting innovation (see point 3).

2.1. A European policy for an information and knowledge-based society

Regarding the demand for knowledge, this policy must:

(a) encourage innovation in products and services with a larger knowledge input which can improve the quality of citizens' lives (in the field of transport, tourism, the environment, public administration, health and assistance to the elderly). This is a wide frontier to be explored with a view to boosting job creation;
(b) speed up the diffusion, in companies, of information technologies linked to flexible production systems, e-commerce, teleworking, telemedicine and also of information and knowledge-management tools;
(c) improve training for workers to help them cope with information technologies by adopting a reference frame of basic skills, setting up a European network of open learning centres equipped with multimedia technologies and distance teaching facilities and encouraging continuing training and the creation of learning organisations in

companies; a 'European passport' for information technologies must be a priority objective. The aim is to stimulate the acquisition and certification of basic IT skills by the entire European population and, to make it compulsory, for future generations of students;

(d) spread information technologies throughout the education and training system by providing all establishments with Internet-linked computers and suitably trained staff;

(e) provide guidance and educational and professional support to encourage everyone to adjust to the new requirements of information and knowledge-based society, giving particular attention to those categories in danger of serious social exclusion;

(f) modernise public services by using information technologies to improve citizens' and companies access to both information and the provision of services.

As regards the supply of knowledge, the policy must:

(a) bolster European R&D networks through closer cooperation and coordination not only under the Framework Programme but also all other programmes for international scientific and technological cooperation (e.g., EUREKA and COST), those of intergovernmental scientific organisations (ESA, CERN, EMBL, ESRF, etc.) as well as national programmes;

(b) develop content industries and set up a content database accessible to the public. Here it will be essential to take measures to promote digitalisation and accessibility of all information of interest to the public, and to make any State-held contents available to industry with a view to fostering added value;

(c) develop software for communicating and generating knowledge (specifically in the field of computational processing of natural languages, in order to boost interchange between languages and cultures, content industries, e-commerce, telematics for educational purposes, etc.);

(d) speed up the construction of trans-European broadband telecommunications networks (liberalisation, definition of standards, interoperability) and promote their accessibility on terms that are internationally competitive.

The e-Europe European initiative recently proposed by the European Commission and the latter's communication on employment strategy in the information society should act as catalysts for the Action Plan to be prepared forthwith by the Presidency and the Commission, as decided at the Helsinki European Council.

The Lisbon European Council will define the Action Plan's objectives and guidelines in order to enable the Presidency and the Commission to work out benchmarking indicators to be included in national initiatives on the information and knowledge-based society to apply from 2001, in accordance with the method of open coordination among Member States to be approved by the Feira European Council in June.

Reference indicators to assist benchmarking schemes for genuine political advances should be developed as a matter of urgency. The adoption of a Plan for an Information and Knowledge-based Society should firmly commit Europe to meeting identifiable targets and disseminating best practices. This European Action Plan should serve as a guide to national plans integrated into national development strategies and linked to national employment schemes.

2.2. Creation of a Europe-wide learning society

European policies on education and training should go beyond the successive reforms of the existing systems already implemented. The aim should be to create a European area of life-long learning and to bring about a learning society with opportunities for all. Without a learning society, the changeover to a knowledge-based economy will cause new breaches and new forms of social exclusion.

A learning society should be geared towards giving diversified and relevant answers to a wide range of target groups: young people, unemployed adults, workers at risk, but also businessmen and middle and senior executives, not to mention the large mass of workers who need to be given genuine life-long training opportunities. This is also a key area for action by the social partners, since there can be no life-long learning without the involvement of enterprises. It should likewise be a central concern of youth policies in the European Union.

Policies on education and training should, moreover, be geared towards creating a large stock of skilled jobs. The potential is already there. However, those jobs will only actually be created if there are skilled human resources to fill them.

The differences between Member States' education and training systems are huge; yet, notwithstanding the more general aims of citizens' personal, social and cultural development, concerns regarding the relevance of training for coping with the requirements of the new jobs are shared by all. Thus, the education and training input for the Luxembourg Process employment guidelines should be reinforced, to meet shared problems, namely by:

(a) developing schools and training centres, into learning centres, using the most appropriate methods to cope with a greater diversity of target groups; promoting cooperation between education and training establishments and putting new learning facilities to good use;

(b) fostering the mobility of students, teachers and training staff, notably through recognition of diplomas and periods of study and training;

(c) equipping the said establishments with computer equipment and Internet connections and staffing them with teachers and other experts with up-to-date training geared to the objectives;

(d) renewing content production in combination with curriculum development and promoting the widespread development of scientific and technical skills as the essential basis for creating skilled employment;

(e) working out educational and vocational guidance schemes of general application, based on the identification of training requirements;

(f) setting up flexible schemes to certify knowledge acquired;

(g) introducing new forms of funding and time management to facilitate access to life-long education and training.

The Socrates and Leonardo da Vinci programmes, which are Commission initiatives, will be launched at the Conference of Ministers for Education, Labour and Social Affairs on 17 and 18 March. These programmes will stimulate the European dimension of education and training through the organisation of exchange networks, the production of common content and reference frames, the identification of training requirements and the promotion of mobility and qualification equivalence. This fundamental action needs to be reinforced; however, it would be even more effective if it were supplemented by a method of open coordination between Member States.

The Lisbon European Council should take forward:

(a) the definition of the contribution of education policies towards employment policies under the Luxembourg Process;

(b) other forms of coordination between Member States, particularly as regards mobility of teachers, students and training staff and the possibility of drawing up a European Charter of Basic Skills, with implications for curriculum updating. Europe's population, and young people in particular, must have extensive access to basic skills, such as being able to learn and to resolve problems, develop scientific culture and technical skills, use information technologies, speak foreign languages, develop a sense of initiative and entrepreneurship and be active, free and responsible citizens.

2.3. Developing a European research area open to the world

European S&T policy has passed through various stages: after focusing on major fundamental research projects at the European level and subsequently on major European pre-competitive projects, it has recently switched to research linked to company innovation. On the whole, and notwithstanding major national differences, comparison between the EU and the US reveals some significant shortcomings as things now stand:

- restrictions on innovative fundamental research;
- reduced sensitivity of public research to market requirements;
- reduced private-sector contribution to S&T activities;
- fragile interface between S&T supply and demand;
- less effort to promote a scientific and technical culture;
- under-developed institutional capacity to conduct European S&T policies.

Yet Europe, having an abundant heritage and capacity, should strive to be in the forefront of the development of scientific knowledge. This is an essential condition for renovating the economic base for employment. An organisational effort is required at two levels:

- fundamental research: here, an added European dimension is needed to achieve scale and scope. This involves organising a European research area using joint activity networks and joint infrastructures in order to overcome the current situation of fragmentation and overlapping of national institutions. The R&D Framework Programme, in coordination with other policy instruments, should serve as a stimulus to such networks;
- applied research and development: here it is necessary to stimulate company-based R&D, to internationalise – but also to develop the interfaces with companies and with regional innovation systems, taking advantage of the diversity of competitive factors in Europe in the case of both the high-tech industries and the traditional industries undergoing modernisation, as well as in the field of services.

The linkage between these two levels of R&D must grasp new opportunities. It is proving necessary nowadays to set up more effective coordination mechanisms in many areas, for example oceanography, meteorology, etc.

The following priorities should be emphasised:

(a)   Stepping up European coordination with regard to the various ways of organising S&T systems, combining local, sectoral and international resources in order to produce, disseminate and adapt knowledge and exploit new opportunities for scientific and technological development, adapting existing coordination bodies or creating appropriate European bodies.

(b)   Creating a high-speed, low-cost trans-European data transmission network to provide support for the construction of a European research area. This broad-band network will have to support not only the development of European S&T cooperation, but also cooperation between schools and training centres, libraries and science centres and museums.

(c)   Creating the conditions for encouraging the mobility of S&T staff and opening up European systems to exchange with the outside world. Europe must strengthen its role as a major world centre for R&D, fully integrated into the big international networks and able to attract new talent from anywhere in the world.

(d)   Promoting the scientific and technological education of European citizens, stimulating cooperation between research institutions and schools, encouraging the production of multilingual scientific works, promoting their export outside Europe and making careers in science and technology more attractive.

On the basis of the Commission communication on a European Research Area, the Lisbon European Council will have to define the medium-term joint European objectives and prepare for decisions on the new European S&T initiatives, as well as ways and means of coordinating national S&T policies.

3.   ECONOMIC REFORMS FOR COMPETITIVENESS AND INNOVATION

The establishment of the single European market has been essential for European construction. The gradual elimination of barriers to the free movement of goods and services, together with competition policy and privatisation procedures, has produced very encouraging results for firms and consumers in various sectors: suffice it to mention the recent case of telecommunications. The economic basis for employment has been comprehensively improved.

But the aforementioned effort has to continue, as indicated in the recently approved strategy for the internal market, in conjunction with the Cardiff process of economic reforms, not only in order to improve how the markets work, but also to create new competitive factors, increase innovation potential and develop entrepreneurship. European markets have to adapt to the prospects opened up by an economy based on innovation and knowledge.

The Lisbon European Council will have to:

- Establish an open method of coordination to provide impetus for the Cardiff process, organising a process of benchmarking centred on sound practices relating to priorities defined at the European level and making use of initiatives launched by the European Commission, such as the action plan for financial services, the action plan for risk capital, the proposal on the European patent and the initiatives relating to the policy of support for enterprises.
- Define guidelines for the elaboration of a European Charter for micro-enterprises for final evaluation at the Feira European Council, with the aim of encouraging this new potential for creating employment.

The following priorities have been identified in the context of the Cardiff process:

(a) to increase the opportunities for trade by developing telecommunications and e-commerce;
(b) to improve transport logistics and the transport network to cope with the increase in trade;
(c) to modernise public services, in particular by using various public–private partnership arrangements;
(d) to accelerate the integration of the financial markets, implementing in full the action plan for financial services;
(e) to improve the sensitivity of the financial markets to the value of intangible investments and investments in knowledge;
(f) to encourage access to risk capital at the European and local levels, entirely in line with the action plan for risk capital;
(g) to establish a single European patent system and organise the technological know-how markets;
(h) to encourage entrepreneurial initiative and development.

Specific action by the Member States and the European Commission is also needed to encourage the networks and dynamics of innovation: entrepreneurial innovation, financial innovation, human resources innovation and more efficient technology-transfer mechanisms. There is a need to signal a new frontier to be explored for entrepreneurial initiative: that of goods and services with greater substance in terms of knowledge, in line with new needs. Many quality jobs can be created through this dynamic.

Another significant impulse for competitiveness and employment will be provided by the definition under elaboration of the policy for support to enterprises with a view to preparation for the new multiannual plan. That policy should place emphasis on the following priorities:

(a) support for the incorporation of new technologies and the creation of intensive knowledge services for support to firms;
(b) development of clusters and innovation networks;
(c) encouragement of partnership and associative relations, both locally and internationally;
(d) organisational innovation and new instruments for knowledge management;
(e) development of certification procedures linked to the promotion of total quality;
(f) adapted financial instruments;
(g) adapted schemes for the training of human resources;
(h) simplification of administrative procedures and modernisation of public support services.

It is also important to see to specific needs in each standard case. For example:

- SMEs of a high technological level give rise to development needs and appropriate procedures for access to the capital market and additional technological capabilities.
- Start-up companies must be stimulated on the basis of strengthened risk capital, technical and logistical support and the simplification of procedures and obligations.
- Besides simplification of procedures and obligations, micro-enterprises may encounter encouraging prospects with the development of e-commerce and the upgrading of local networks and entities. This new potential to create micro-enterprises and proliferate small-scale entrepreneurial initiative should be given impetus by the European charter for micro-entreprises.

4. RENEWING THE EUROPEAN SOCIAL MODEL: MORE JOBS AND GREATER SOCIAL COHESION

The European Union has already defined, notably in the context of the European employment strategy, the major priorities for the employment objective. The twofold strategic objective of combating unemployment and increasing the employment rate requires:

- the creation of jobs in the services sector, in which Europe has significant deficits and opportunities for growth;
- a resolute inversion of the trend towards early retirement from the labour market, promoting the employment of older workers;
- an increase in the rate of female employment, encouraging equal opportunity of access to the labour market and positive action in favour of the employment of women.

To make use of the above employment potential also requires stock to be taken of the European social model, which is one of the strong suits of the European project. But there are two prerequisites for its continuation in the context of globalisation: the renovation of its economic base, building new competitive factors, and the modernisation of its very structure. This will make it possible to find a new synthesis with more jobs and greater social cohesion.

Understanding the contemporary problems of the European social model and finding solutions to them requires starting from an adequate concept of welfare. Welfare is not only a guarantee of income in the face of social risks. Welfare is also based on personal services, quality of work and living opportunities. And all this contributes to the cohesion of a society.

A positive strategy of renewal of the European social model needs to be adopted:

- aimed at raising employment levels and creating job opportunities for all;
- combining the principles of initiative, responsibility, social justice and solidarity.

The above renewal endeavour should involve not only the official authorities, but also the remaining protagonists at the various levels, with the accent on the social partners and the role of the European social dialogue. The contributions secured in this way could be used to develop a European social agenda, along lines to be defined during the French Presidency.

Apart from reconsideration of the concepts of employment, work and activity, new ways of regulating the labour market will have to be developed, combining flexibility and security, an area which will require a significant contribution from the social partners. In addition to the progress made in the field of working conditions and minimum standards, there is a need to:

- strengthen the role of active employment policies;
- modernise social protection systems, consolidating their sustainability;
- increase the efficacy of policies to combat social exclusion, which should ensure a solution when the foregoing fail.

4.1. Active policies and the European employment strategy

The Luxembourg process has made a significant contribution to strengthening active employment policies in the framework of the European employment strategy.

A mid-term review of the Luxembourg process will be carried out during the Portuguese Presidency. That review should contribute to a major rationalisation of the guidelines and to more detailed indicators, but it should also coincide with strategic discussions. It is not a matter of revising the guidelines, but first of examining the synergies between these guidelines and between employment guidelines and broad economic guidelines. It is therefore a matter of harmonising the methodology for the involvement of other protagonists, in particular the social partners.

The Lisbon European Council will have to lay down orientations for the aforementioned mid-term review.

From the point of view of creating employment and renewing the European social model, four areas warrant priority:

(a) Improving the efficacy of active employment policies as regards employability. Guaranteeing a prompt response, using information technologies in employment services, diversifying alternatives and making full use of the local level are basic elements for success. It is particularly important to agree on programmes which enable the direct conversion of the unemployed through qualifications which are in shortage on the labour market.

(b) Strengthening the synergies between adaptability and lifelong learning. The development of learning organisations, recourse to the ongoing training network, management of working time, job rotation and cost-sharing should be central to the redefinition of the labour contract and to negotiations between social partners at all levels. The official authorities should also contribute, with specific support, to facilitating such negotiations. Firms which invest in their human resources should benefit from fiscal and parafiscal incentives.

(c) Increasing employment in services, facilitating entrepreneurial initiative, decreasing the administrative burden on SMEs, reducing the non-wage costs of the less qualified and encouraging equal opportunities. As regards personal services, where there are major shortages, private, public or third-sector initiatives can be associated, possibly using a social voucher in favour of the least-favoured categories.

(d) Developing mainstreaming to promote equal opportunities, with particular implications for all aspects which help to reconcile working life and family life. Strengthening the family-support services, in particular child care services, is of special importance.

4.2. Modernising social protection, consolidating its sustainability

Where the population is ageing, where new forms of family are emerging, where new risks on the labour market are surfacing, the tension arising therefrom tends to centre on social security benefits schemes, in particular pensions. This trend is becoming burdensome and is today of concern to all European governments, irrespective of the great diversity of social protection schemes. For that reason, a cooperation process at the European level was recently initiated for the modernisation of those schemes. Taking into account the recent communication from the European Commission on a concerted strategy for modernising social protection, the High-Level Working Party now set up will have to opt for evaluation of the long-term sustainability of those schemes as a priority.

Under the present circumstances, bolstering the sustainability of protection schemes depends to a large measure on one factor: increasing the rate of employment of European populations, which is at a particularly low level. The rate of employment in the European Union at the end of the 1990s was little more than 60%, as against figures in excess of 75% in the USA and Japan.

An increase in the rate of employment implies a considerable improvement of the net creation of jobs. To achieve that goal, in addition to the need to ensure macro-economic stability and galvanise growth factors, it is necessary to improve the actual operation of the labour market, in particular by:

(a) strengthening employability and adaptability on the basis of lifelong education to prevent unemployment;
(b) increasing the efficiency of active employment policies, activating social policies;
(c) modernising fiscal and parafiscal systems so that employment is of benefit to all citizens;
(d) promoting more active ageing, combating early retirement from the labour market;
(e) making working time flexible throughout working life, enabling a

better balance between working life and family life and more flexible careers based on upholding basic social-protection and access-to-training rights.

But consolidation of the sustainability of social benefits is not the only issue. Welfare is not only guaranteed income. It is also access to services. The crisis for the traditional family unit needs to be offset by the development of family-support services, especially for children and the elderly. A large range of services should therefore be encouraged for that purpose, with the advantage of also being very intensive in the creation of new jobs.

The above therefore seem to be some of the basic components for a positive strategy for the modernisation of social protection and the renovation of the European social model. It is a positive strategy because:

- it bolsters the sustainability of social protection schemes;
- it strengthens support for families;
- it reinforces equality of opportunities for men and women;
- by introducing greater flexibility, it maintains basic security and opens new prospects for upward mobility in the labour market;
- it is based on the creation of more jobs.

That strategy will improve the overall outcome for welfare and social cohesion, in line with the broad concept referred to above.

The Lisbon European Council will:

(a) approve the setting up of a high-level working party on the modernisation of social protection and will define its working priorities, with emphasis on the carrying out of a forecast study on the sustainability of the pensions scheme for the period 2010–20;
(b) define the forms of cooperation and of exchanges of best practices between Member States;
(c) call upon the European Commission to develop the procedures necessary for strengthening information systems on social protection.

4.3. Stepping up the fight against social exclusion

Europe in the twenty-first century needs to have a systematic policy to combat poverty and social exclusion in their old and new forms.

In spite of the high levels of economic development of the EU as a whole and the existence of significant social protection instruments, social exclusion still abounds in various forms.

The above phenomenon affects all Member States, albeit in various forms and intensities, in particular the most vulnerable social groups, the most deprived economic areas and those citizens who are particularly disadvantaged as regards the labour market.

On the other hand, social dynamics continue to give rise, with some frequency, to the emergence of child poverty and social integration problems for children and young people.

The intensity of the changes which are foreseen from the point of view of the qualifications required by new technological challenges facing firms also involve the risk of developing new social exclusion processes.

The problem of social exclusion therefore requires major coordination at European level.

On the basis of the report on the social situation and of the Commission communication 'Towards a Europe for all', the Lisbon European Council will define the open method of coordination which will have to be applied respectively to two forms of action to be combined by each Member State, involving the other active partners in promoting social inclusion:

(a) to mainstream this objective in education, training, employment and social protection policies;
(b) to develop integrated, targeted programmes for social groups in situations of major social exclusion, with the top priority of eradicating child poverty by 2010.

The Lisbon European Council will also call upon the High-Level Working Party and the European Commission to prepare a monitoring panel with indicators for monitoring the social situation, making it possible to set policy objectives which can draw on the experience of the various Member States.

5. MACRO-ECONOMIC POLICIES FOR SUSTAINABLE GROWTH

There is one central item on the European political agenda: achieving a policy mix which stimulates growth and employment, ensuring macro-economic stability and consolidation of the euro. This means, in the context of the Stability and Growth Pact, stimulating growth and the transition to an economy of innovation and knowledge. This will in particular mean assigning a more important role to structural policies and reforms.

A fundamental pillar of the policy mix is monetary and exchange policy. The main aim of monetary policy, defined and implemented by the Eurosystem, is to guarantee price stability. The extent to which it can support the Community's general economic policies depends on what happens in the other fundamental pillars of the policy mix. The second pillar is fiscal policy defined and implemented by national governments, in the light of the provisions of the Stability and Growth Pact. The third pillar consists of wage developments, chiefly determined by negotiations between the social partners.

It is important to continue to monitor budgetary policies on many fronts, in particular endeavouring to:

(a) Improve methods of monitoring expenditure, debt and deficit, as regards not only level, but also content.

(b) Adjust the monitoring of deficits so as to maximise the room for manoeuvre of automatic stabilisers.

(c) Improve the quality of public spending, redirecting it towards promoting public investment and meeting new priorities (R&D, education and training, modernisation of the public administration, affordable telecommunications, etc.).

(d) Define new methods of public–private partnership so as to speed up the investment necessary to the modernisation of economies, while also making use of new products to be launched on the financial markets.

(Specifically with regard to tax policy):

(a) Develop tax coordination, endeavouring to overcome problems of harmful competition as yet unresolved.

(b) Make progress as regards changes in taxation more friendly to the aims of employment and social cohesion.

(c) Reinforce those tax incentives which encourage firms and individuals to adapt to an economy of innovation and knowledge.

(d) Promote a tax system more favourable to SMEs.

However, it is also vital to improve multilateral coordination of macroeconomic policies so as to make full use of Economic and Monetary Union to encourage growth on a sustainable basis.

Economic coordination may take place in successive stages. Some stages are already taking place: the setting of common objectives, their translation into national plans and multilateral monitoring of their implementation. However, it is possible to move on to further levels of coordination:

(a) Evaluation of the aggregate effects of the various choices made at the national level.

(b) The development of strategies for coping with problems such as asymmetric or global shocks and the creation of large-scale infrastructure at the European level. In order to exercise a greater effect on such large-scale infrastructure, we must consider making greater use of the role of the EIB and the EIF, of public–private partnerships and of other financial instruments, and strengthening the trans-European networks programme, particularly with regard to 'knowledge structures'.

On the other hand, it is vital that we define and coordinate more clearly the role of each of the bodies and protagonists involved: ECOFIN, Euro-11, EFC, SCE, ECB, the social partners, macro-economic dialogue. The aim is to create a relationship of trust and interchange between all the parties involved, and in this context macro-economic dialogue may play an important role.

6. METHODS FOR ACHIEVING A EUROPEAN DIMENSION

The political construction of Europe is a unique experience. Its success has been dependent on the ability to combine coherence with respect for diversity and efficiency with democratic legitimacy. This entails using different political methods depending on policies and the various institutional processes. For good reasons, various methods have been worked out which are placed somewhere between pure integration and straightforward cooperation. Hence:

- Monetary policy is a single, common policy within the euro zone.
- National budgetary policies are coordinated at the European level on the basis of strictly predefined criteria.
- Employment policies are coordinated at the European level on the basis of guidelines and certain indicators, allowing some room for adjustment at the national level.
- A process of cooperation is beginning with a view to the modernisation of social protection policies, with due regard for national differences.

Policies aimed at building the single market, such as monetary policy or competition policy are based, as is logical, on a stricter method of coordination as regards the principles to be observed. However, there are other policies which concentrate more on creating new skills and capacities for making use of this market and responding to structural changes. They involve learning more quickly and discovering appropriate solutions. Such policies have resulted in the formulation of a coordination method which is more open to national diversity, the best example of which currently is the so-called 'Luxembourg process' relating to employment policies.

It is a case of defining strategic guidelines at the European level for coping with structural change and then organising a process whereby Member States emulate each other in applying them, stimulating the exchange of best practices, while taking account of national characteristics. Despite some difficulties, the results obtained have been stimulating and encouraging.

The open method of coordination varies in intensity depending on the subject areas to which it is applied and on how the subsidiarity principle is expressed in each of them. In its most complete form, this open method of coordination consists of the following steps:

(a)   In the light of diagnosis and evaluation, setting of Europe-wide guidelines with the political commitment to apply them being assumed at the highest level.

(b)   Identifying good practices and reference indicators for benchmarking purposes in these guidelines.

(c)   Preparing national plans for applying these guidelines in a suitable way so as to involve all the various protagonists, identifying intermediate goals and learning processes.

(d)   Organising the various partnerships responsible and implementing the national plan.

(e)   Monitoring and evaluating the results obtained, allowing for discussion and peer pressure and possibly formulation of recommendations.

In this context, the European Commission initiative programmes may gain additional scope and effectiveness. As well as promoting a typically European dimension, as is their aim, they may play an extremely important role in supporting the whole process of the open method of coordination between Member States. Thus, it will be possible to step up the efforts made on the basis of the Community budget and the efforts made by the Member States, depending on their own circumstances.

Here the European Parliament should also be encouraged to become involved and the other European Union institutions consulted.

If economic and social innovation is to be stimulated, there is also a need for innovation in the political method. This open method of coordination also makes it possible to progress with due regard for diversity. It will have to be applied with the necessary adjustments in new areas, as proposed in the previous sections.

In order to ensure that there is overall coherence and that objectives are the same in all areas in which this method is applied, the Lisbon European Council will have to ask the European Commission to draw up a proposal for a monitoring panel of the most important indicators for structural change, showing their effect on the rate of economic growth and the rate of employment throughout the European Union.

7. COORDINATION OF POLICIES FOR A EUROPEAN STRATEGY

A European growth and employment strategy requires better coordination between macro-economic policies, structural policies and reforms and active employment policies, based on the Cologne, Cardiff and Luxembourg processes.

These processes overlapped in time and their procedures and timetables arose from their beginnings and specific motivations. Taking into account the framework set by the Treaties and the different levels of subsidiarity and specialised Councils they involve, there is justification for preserving them as three distinct processes. However, the time now seems to have come for their coordination, synergy and joint efficiency to be improved.

The Lisbon European Council will have to develop the conclusions adopted at the Helsinki European Council so that the Broad Economic Policy Guidelines (BEPG) carry even more weight as a framework document, as regards not only the various macro-economic policies, but also political and structural reforms, and also the connections with the employment guidelines. The coherence and synergy between these three components must be dealt with explicitly and systematically.

For that purpose, ECOFIN will have to receive contributions from other Council formations, particularly from Labour and Social Affairs, but also namely from the Internal Market and Industry, as part of a coordination process subject to the political guidance of the European Council. The Cardiff and Luxembourg processes will make it possible for us to deal with their subject matter in greater detail.

In addition, the BEPG must define the guidelines to be adopted by the EU and the recommendations to the Member States, providing a framework for next year and looking forward to the years ahead, while taking into account recent developments in the Member States.

The importance of the BEPG justifies more substantial involvement by the European Council in their general drafting.

More political weight should also be given to the mechanisms for monitoring the formulation and implementation of recommendations, with the involvement of the relevant specialised Councils.

On the other hand, it will be important to inform and consult the social partners on the basis of the structures provided for in the social dialogue, the Standing Committee on Employment and the macro-economic dialogue, so as to identify the contribution they can make to the guidelines to be adopted and to the implementation of the European Employment Pact.

It will be for the Portuguese Presidency to conduct the first practical exercise in applying this new concept of the BEPG, in terms to be defined by the Lisbon European Council.

Another coordination instrument which must be given more weight is the annual report prepared by the European Commission based on Article 127 of the Treaty, on 'Community policies in support of employment', which explains how to use mainstreaming to achieve the aim of employment on the basis of Community policies.

The setting up of an Observatory on Industrial Change on a proposal from the European Commission to be approved by the Lisbon European Council must also reinforce the exchange of best practices for the management of change, with the involvement of the various actors, in particular the social partners and enterprises in general.

The quest for concerted action must involve many other actors in addition to governments and the European Commission, starting with the major institutions such as the European Parliament and national parliaments, the European Central Bank, the social partners, the Economic and Social Committee and the Committee of the Regions. A High-Level Forum will take place before the June European Council in order to review the various processes and share out responsibility for directing them. The aim is to enhance the content of the European Employment Pact adopted in Cologne. The Lisbon European Council will consider the possibility of holding this *Forum* annually.

Finally, such action depends to a large extent on the initiative of the actors in civil society, the social partners, enterprises, associations, regions and the citizens in a European civil society, which we must continue to build.

Europe's capacity to influence its own mode of development depends on all of the above. The Portuguese Presidency is counting on the participation and commitment of all these actors in order to give a new long-term impetus to the construction of Europe. The time has come.

Presidency conclusions

Lisbon European Council 23 and 24 March 2000
(relevant paragraphs)

The European Council held a special meeting on 23–24 March 2000 in Lisbon to agree a new strategic goal for the Union in order to strengthen employment, economic reform and social cohesion as part of a knowledge-based economy. At the start of proceedings, an exchange of views was conducted with the President of the European Parliament, Mrs Nicole Fontaine, on the main topics for discussion.

I. EMPLOYMENT, ECONOMIC REFORM AND SOCIAL COHESION

A strategic goal for the next decade

The new challenge

1. The European Union is confronted with a quantum shift resulting from globalisation and the challenges of a new knowledge-driven economy. These changes are affecting every aspect of people's lives and require a radical transformation of the European economy. The Union must shape these changes in a manner consistent with its values and concepts of society and also with a view to the forthcoming enlargement.

2. The rapid and accelerating pace of change means it is urgent for the Union to act now to harness the full benefits of the opportunities presented. Hence the need for the Union to set a clear strategic goal and agree a challenging programme for building knowledge infrastructures, enhancing innovation and economic reform, and modernising social welfare and education systems.

The Union's strengths and weaknesses

3. The Union is experiencing its best macro-economic outlook for a generation. As a result of stability-oriented monetary policy supported

by sound fiscal policies in a context of wage moderation, inflation and interest rates are low, public sector deficits have been reduced remarkably and the EU's balance of payments is healthy. The euro has been successfully introduced and is delivering the expected benefits for the European economy. The internal market is largely complete and is yielding tangible benefits for consumers and businesses alike. The forthcoming enlargement will create new opportunities for growth and employment. The Union possesses a generally well-educated workforce as well as social protection systems able to provide, beyond their intrinsic value, the stable framework required for managing the structural changes involved in moving towards a knowledge-based society. Growth and job creation have resumed.

4.   These strengths should not distract our attention from a number of weaknesses. More than 15 million Europeans are still out of work. The employment rate is too low and is characterised by insufficient participation in the labour market by women and older workers. Long-term structural unemployment and marked regional unemployment imbalances remain endemic in parts of the Union. The services sector is underdeveloped, particularly in the areas of telecommunications and the Internet. There is a widening skills gap, especially in information technology where increasing numbers of jobs remain unfilled. With the current improved economic situation, the time is right to undertake both economic and social reforms as part of a positive strategy which combines competitiveness and social cohesion.

The way forward

5.   The Union has today set itself a ***new strategic goal*** for the next decade: *to become the most competitive and dynamic knowledge-based economy in the world capable of sustainable economic growth with more and better jobs and greater social cohesion*. Achieving this goal requires an ***overall strategy*** aimed at:

- preparing the transition to a knowledge-based economy and society by better policies for the information society and R&D, as well as by stepping up the process of structural reform for competitiveness and innovation and by completing the internal market;
- modernising the European social model, investing in people and combating social exclusion;
- sustaining the healthy economic outlook and favourable growth prospects by applying an appropriate macro-economic policy mix.

6. This strategy is designed to enable the Union to regain the conditions for full employment, and to strengthen regional cohesion in the European Union. The European Council needs to set a goal for full employment in Europe in an emerging new society which is more adapted to the personal choices of women and men. If the measures set out below are implemented against a sound macro-economic background, an average economic growth rate of around 3% should be a realistic prospect for the coming years.

7. Implementing this strategy will be achieved by improving the existing processes, introducing a *new open method of coordination* at all levels, coupled with a stronger guiding and coordinating role for the European Council to ensure more coherent strategic direction and effective monitoring of progress. A meeting of the European Council to be held every Spring will define the relevant mandates and ensure that they are followed up.

PREPARING THE TRANSITION TO A COMPETITIVE, DYNAMIC AND KNOWLEDGE-BASED ECONOMY

An information society for all

8. The shift to a digital, knowledge-based economy, prompted by new goods and services, will be a powerful engine for growth, competitiveness and jobs. In addition, it will be capable of improving citizens' quality of life and the environment. To make the most of this opportunity, the Council and the Commission are invited to draw up a comprehensive *e*Europe Action Plan to be presented to the European Council in June this year, using an open method of coordination based on the benchmarking of national initiatives, combined with the Commission's recent *e*Europe initiative as well as its communication 'Strategies for jobs in the Information Society'.

9. Businesses and citizens must have access to an inexpensive, world-class communications infrastructure and a wide range of services. Every citizen must be equipped with the skills needed to live and work in this new information society. Different means of access must prevent info-exclusion. The combat against illiteracy must be reinforced. Special attention must be given to disabled people. Information technologies can be used to renew urban and regional development and promote environmentally sound technologies. Content industries create added value by exploiting and networking

European cultural diversity. Real efforts must be made by public administrations at all levels to exploit new technologies to make information as accessible as possible.

10. Realising Europe's full e-potential depends on creating the conditions for electronic commerce and the Internet to flourish, so that the Union can catch up with its competitors by hooking up many more businesses and homes to the Internet via fast connections. The rules for electronic commerce must be predictable and inspire business and consumer confidence. Steps must be taken to ensure that Europe maintains its lead in key technology areas such as mobile communications. The speed of technological change may require new and more flexible regulatory approaches in the future.

11. The European Council calls in particular on:

- the Council along with the European Parliament, where appropriate, to adopt as rapidly as possible during 2000 pending legislation on the legal framework for electronic commerce, on copyright and related rights, on e-money, on the distance selling of financial services, on jurisdiction and the enforcement of judgements, and the dual-use export control regime; the Commission and the Council to consider how to promote consumer confidence in electronic commerce, in particular through alternative dispute resolution systems;
- the Council and the European Parliament to conclude as early as possible in 2001 work on the legislative proposals announced by the Commission following its 1999 review of the telecoms regulatory framework; the Member States and, where appropriate, the Community to ensure that the frequency requirements for future mobile communications systems are met in a timely and efficient manner. Fully integrated and liberalised telecommunications markets should be completed by the end of 2001;
- the Member States, together with the Commission, to work towards introducing greater competition in local access networks before the end of 2000 and unbundling the local loop in order to help bring about a substantial reduction in the costs of using the Internet;
- the Member States to ensure that all schools in the Union have access to the Internet and multimedia resources by the end of 2001, and that all the teachers needed are skilled in the use of the Internet and multimedia resources by the end of 2002;
- the Member States to ensure generalised electronic access to main basic public services by 2003;

- the Community and the Member States, with the support of the EIB, to make available in all European countries low-cost, high-speed interconnected networks for Internet access and foster the development of state-of-the-art information technology and other telecom networks as well as the content for those networks. Specific targets should be defined in the *e*Europe Action Plan.

Establishing a European Area of Research and Innovation

12. Given the significant role played by research and development in generating economic growth, employment and social cohesion, the Union must work towards the objectives set out in the Commission's communication 'Towards a European Research Area'. Research activities at the national and Union level must be better integrated and coordinated to make them as efficient and innovative as possible, and to ensure that Europe offers attractive prospects to its best brains. The instruments under the Treaty and all other appropriate means, including voluntary arrangements, must be fully exploited to achieve this objective in a flexible, decentralised and non-bureaucratic manner. At the same time, innovation and ideas must be adequately rewarded within the new knowledge-based economy, particularly through patent protection.

13. The European Council asks the Council and the Commission, together with the Member States where appropriate, to take the necessary steps as part of the establishment of a European Research Area to:

  - develop appropriate mechanisms for networking national and joint research programmes on a voluntary basis around freely chosen objectives, in order to take greater advantage of the concerted resources devoted to R&D in the Member States, and ensure regular reporting to the Council on the progress achieved; to map by 2001 research and development excellence in all Member States in order to foster the dissemination of excellence;
  - improve the environment for private research investment, R&D partnerships and high technology start-ups, by using tax policies, venture capital and EIB support;
  - encourage the development of an open method of coordination for benchmarking national research and development policies and identify, by June 2000, indicators for assessing performance in different fields, in particular with regard to the development

of human resources; introduce by June 2001 a European innovation scoreboard;

- facilitate the creation by the end of 2001 of a very high-speed
transeuropean network for electronic scientific communications, with EIB support, linking research institutions and universities, as well as scientific libraries, scientific centres and,
progressively, schools;
- take steps to remove obstacles to the mobility of researchers in
Europe by 2002 and to attract and retain high-quality research
talent in Europe;
- ensure that a Community patent is available by the end of 2001,
including the utility model, so that Community-wide patent
protection in the Union is as simple and inexpensive to obtain
and as comprehensive in its scope as the protection granted by
key competitors.

**Creating a friendly environment for starting up and developing innovative
businesses, especially SMEs**

14. The competitiveness and dynamism of businesses are directly dependent on a regulatory climate conducive to investment, innovation and
entrepreneurship. Further efforts are required to lower the costs of
doing business and remove unnecessary red tape, both of which are
particularly burdensome for SMEs. The European institutions,
national governments and regional and local authorities must continue to pay particular attention to the impact and compliance costs
of proposed regulations, and should pursue their dialogue with business and citizens with this aim in mind. Specific action is also needed
to encourage the key interfaces in innovation networks, i.e. interfaces
between companies and financial markets, R&D and training institutions, advisory services and technological markets.

15. The European Council considers that an open method of coordination should be applied in this area and consequently asks:

- the Council and the Commission to launch, by June 2000, a
benchmarking exercise on issues such as the length of time and
the costs involved in setting up a company, the amount of risk
capital invested, the numbers of business and scientific graduates and training opportunities. The first results of this exercise
should be presented by December 2000;
- the Commission to present shortly a communication on an
entrepreneurial, innovative and open Europe together with the
Multiannual Programme in favour of Enterprise and

Entrepreneurship for 2001–2005, which will play an important role as catalyst for this exercise;

- the Council and the Commission to draw up a European Charter for small companies to be endorsed in June 2000 which should commit Member States to focus in the abovementioned instruments on small companies as the main engines for job-creation in Europe, and to respond specifically to their needs;
- the Council and the Commission to report by the end of 2000 on the ongoing review of EIB and EIF financial instruments in order to redirect funding towards support for business start-ups, high-tech firms and micro-enterprises, as well as other risk-capital initiatives proposed by the EIB.

Economic reforms for a complete and fully operational internal market

16. Rapid work is required in order to complete the internal market in certain sectors and to improve under-performance in others in order to ensure the interests of business and consumers. An effective frame-work for ongoing review and improvement, based on the Internal Market Strategy endorsed by the Helsinki European Council, is also essential if the full benefits of market liberalisation are to be reaped. Moreover, fair and uniformly applied competition and state aid rules are essential for ensuring that businesses can thrive and operate effectively on a level playing field in the internal market.

17. The European Council accordingly asks the Commission, the Council and the Member States, each in accordance with their respective powers:
- to set out by the end of 2000 a strategy for the removal of barriers to services;
- to speed up liberalisation in areas such as gas, electricity, postal services and transport. Similarly, regarding the use and management of airspace, the Council asks the Commission to put forward its proposals as soon as possible. The aim is to achieve a fully operational internal market in these areas; the European Council will assess progress achieved when it meets next Spring on the basis of a Commission report and appropriate proposals;
- to conclude work in good time on the forthcoming proposals to update public procurement rules, in particular to make them accessible to SMEs, in order to allow the new rules to enter into force by 2002;
- to take the necessary steps to ensure that it is possible by 2003 for Community and government procurement to take place on-line;

- to set out by 2001 a strategy for further coordinated action to simplify the regulatory environment, including the performance of public administration, at both the national and Community level. This should include identifying areas where further action is required by Member States to rationalise the transposition of Community legislation into national law;
- to further their efforts to promote competition and reduce the general level of State aids, shifting the emphasis from supporting individual companies or sectors towards tackling horizontal objectives of Community interest, such as employment, regional development, environment and training or research.

18. Comprehensive structural improvements are essential to meet ambitious targets for growth, employment and social inclusion. Key areas have already been identified by the Council to be reinforced in the Cardiff process. The European Council accordingly invites the Council to step up work on structural performance indicators and to report by the end of 2000.

19. The European Council considers it essential that, in the framework of the internal market and of a knowledge-based economy, full account is taken of the Treaty provisions relating to services of general economic interest, and to the undertakings entrusted with operating such services. It asks the Commission to update its 1996 communication based on the Treaty.

Efficient and integrated financial markets

20. Efficient and transparent financial markets foster growth and employment by better allocation of capital and reducing its cost. They therefore play an essential role in fuelling new ideas, supporting entrepreneurial culture and promoting access to and use of new technologies. It is essential to exploit the potential of the euro to push forward the integration of EU financial markets. Furthermore, efficient risk capital markets play a major role in innovative high-growth SMEs and the creation of new and sustainable jobs.

21. To accelerate completion of the internal market for financial services, steps should be taken:

- to set a tight timetable so that the Financial Services Action Plan is implemented by 2005, taking into account priority action areas such as: facilitating the widest possible access to investment capital on an EU-wide basis, including for SMEs, by means of a 'single passport' for issuers; facilitating the successful participation of all investors in an integrated market eliminating barriers to investment in pension funds; promoting further integration and better functioning of government bond markets through greater consultation and transparency on debt-issuing calendars, techniques and instruments, and improved functioning of cross-border sale and repurchase ('repo') markets; enhancing the comparability of companies' financial statements; and more intensive cooperation by EU financial market regulators;

- to ensure full implementation of the Risk Capital Action Plan by 2003;
- to make rapid progress on the long-standing proposals on take-over bids and on the restructuring and winding-up of credit institutions and insurance companies in order to improve the functioning and stability of the European financial market;
- to conclude, in line with the Helsinki European Council conclusions, the pending tax package.

Coordinating macro-economic policies: fiscal consolidation, quality and sustainability of public finances

22. As well as preserving macro-economic stability and stimulating growth and employment, macro-economic policies should foster the transition towards a knowledge-based economy, which implies an enhanced role for structural policies. The macro-economic dialogue under the Cologne process must create a relationship of trust between all the actors involved in order to have a proper understanding of each other's positions and constraints. The opportunity provided by growth must be used to pursue fiscal consolidation more actively and to improve the quality and sustainability of public finances.

23. The European Council requests the Council and the Commission, using the existing procedures, to present a report by Spring 2001 assessing the contribution of public finances to growth and employment, and assessing, on the basis of comparable data and indicators, whether adequate concrete measures are being taken in order to:

- alleviate the tax pressure on labour and especially on the relatively unskilled and low-paid, improve the employment and training incentive effects of tax and benefit systems;
- redirect public expenditure towards increasing the relative importance of capital accumulation – both physical and human – and support research and development, innovation and information technologies;
- ensure the long-term sustainability of public finances,

examining the different dimensions involved, including the impact of ageing populations, in the light of the report to be prepared by the High Level Working Party on Social Protection.

MODERNISING THE EUROPEAN SOCIAL MODEL BY INVESTING IN PEOPLE AND BUILDING AN ACTIVE WELFARE STATE

24. People are Europe's main asset and should be the focal point of the Union's policies. Investing in people and developing an active and dynamic welfare state will be crucial both to Europe's place in the knowledge economy and for ensuring that the emergence of this new economy does not compound the existing social problems of unemployment, social exclusion and poverty.

Education and training for living and working in the knowledge society
25. Europe's education and training systems need to adapt both to the demands of the knowledge society and to the need for an improved level and quality of employment. They will have to offer learning and training opportunities tailored to target groups at different stages of their lives: young people, unemployed adults and those in employment who are at risk of seeing their skills overtaken by rapid change. This new approach should have three main components: the development of local learning centres, the promotion of new basic skills, in particular in the information technologies, and increased transparency of qualifications.

26. The European Council accordingly calls upon the Member States, in line with their constitutional rules, the Council and the Commission to take the necessary steps within their areas of competence to meet the following targets:
    - a substantial annual increase in per capita investment in human resources;
    - the number of 18 to 24 year olds with only lower-secondary-level education who are not in further education and training should be halved by 2010;
    - schools and training centres, all linked to the Internet, should be developed into multi-purpose local learning centres accessible to all, using the most appropriate methods to address a wide range of target groups; learning partnerships should be established between schools, training centres, firms and research facilities for their mutual benefit;

- a European framework should define the new basic skills to be provided through lifelong learning: IT skills, foreign languages, technological culture, entrepreneurship and social skills; a European diploma for basic IT skills, with decentralised certification procedures, should be established in order to promote digital literacy throughout the Union;
- define, by the end of 2000, the means for fostering the mobility of students, teachers and training and research staff both through making the best use of existing Community programmes (Socrates, Leonardo, Youth), by removing obstacles and through greater transparency in the recognition of qualifications and periods of study and training; to take steps to remove obstacles to teachers' mobility by 2002 and to attract high-quality teachers.
- a common European format should be developed for curricula vitae, to be used on a voluntary basis in order to facilitate mobility by helping the assessment of knowledge acquired, both by education and training establishments and by employers.

27.   The European Council asks the Council (Education) to undertake a general reflection on the concrete future objectives of education systems, focusing on common concerns and priorities while respecting national diversity, with a view to contributing to the Luxembourg and Cardiff processes and presenting a broader report to the European Council in the Spring of 2001.

More and better jobs for Europe: developing an active employment policy

28.   The Luxembourg process, based on drawing up employment guidelines at Community level and translating them into National Employment Action Plans, has enabled Europe to substantially reduce unemployment. The mid-term review should give a new impetus to this process by enriching the guidelines and giving them more concrete targets by establishing closer links with other relevant policy areas and by defining more effective procedures for involving the different actors. The social partners need to be more closely involved in drawing up, implementing and following up the appropriate guidelines.

29.   In this context, the Council and the Commission are invited to address the following four key areas:

- improving employability and reducing skills gaps, in particular by providing employment services with a Europe-wide data

base on jobs and learning opportunities; promoting special programmes to enable unemployed people to fill skill gaps;

- giving higher priority to lifelong learning as a basic component
of the European social model, including by encouraging agreements between the social partners on innovation and lifelong
learning; by exploiting the complementarity between lifelong
learning and adaptability through flexible management of
working time and job rotation; and by introducing a European
award for particularly progressive firms. Progress towards these
goals should be benchmarked;
- increasing employment in services, including personal services,
where there are major shortages; private, public or third sector
initiatives may be involved, with appropriate solutions for the
least-favoured categories;
- furthering all aspects of equal opportunities, including reducing occupational segregation, and making it easier to reconcile
working life and family life, in particular by setting a new
benchmark for improved childcare provision.

30. The European Council considers that the overall aim of these measures should be, on the basis of the available statistics, to raise the
employment rate from an average of 61% today to as close as possible
to 70% by 2010 and to increase the number of women in employment
from an average of 51% today to more than 60% by 2010. Recognising
their different starting points, Member States should consider setting
national targets for an increased employment rate. This, by enlarging
the labour force, will reinforce the sustainability of social protection
systems.

Modernising social protection
31. The European social model, with its developed systems of social protection, must underpin the transformation to the knowledge
economy. However, these systems need to be adapted as part of an
active welfare state to ensure that work pays, to secure their long-term
sustainability in the face of an ageing population, to promote social
inclusion and gender equality, and to provide quality health services.
Conscious that the challenge can be better addressed as part of a
cooperative effort, the European Council invites the Council to:

- strengthen cooperation between Member States by exchanging
experiences and best practice on the basis of improved information networks which are the basic tools in this field;

- mandate the High Level Working Party on Social Protection, taking into consideration the work being done by the Economic Policy Committee, to support this cooperation and, as its first priority, to prepare, on the basis of a Commission communication, a study on the future evolution of social protection from a long-term point of view, giving particular attention to the sustainability of pensions systems in different time frameworks up to 2020 and beyond, where necessary. A progress report should be available by December 2000.

Promoting social inclusion

32. The number of people living below the poverty line and in social exclusion in the Union is unacceptable. Steps must be taken to make a decisive impact on the eradication of poverty by setting adequate targets to be agreed by the Council by the end of the year. The High Level Working Party on Social Protection will be involved in this work. The new knowledge-based society offers tremendous potential for reducing social exclusion, both by creating the economic conditions for greater prosperity through higher levels of growth and employment, and by opening up new ways of participating in society. At the same time, it brings a risk of an ever-widening gap between those who have access to the new knowledge, and those who are excluded. To avoid this risk and maximise this new potential, efforts must be made to improve skills, promote wider access to knowledge and opportunity and fight unemployment: the best safeguard against social exclusion is a job. Policies for combating social exclusion should be based on an open method of coordination combining national action plans and a Commission initiative for cooperation in this field to be presented by June 2000.

33. In particular, the European Council invites the Council and the Commission to:

    - promote a better understanding of social exclusion through continued dialogue and exchanges of information and best practice, on the basis of commonly agreed indicators; the High Level Working Party on Social Protection will be involved in establishing these indicators;
    - mainstream the promotion of inclusion in Member States' employment, education and training, health and housing policies, this being complemented at Community level by action under the Structural Funds within the present budgetary framework;

- develop priority actions addressed to specific target groups (for example minority groups, children, the elderly and the disabled), with Member States choosing amongst those actions according to their particular situations and reporting subsequently on their implementation.

34. Taking account of the present conclusions, the Council will pursue its reflection on the future direction of social policy on the basis of a Commission communication, with a view to reaching agreement on a European Social Agenda at the Nice European Council in December, including the initiatives of the different partners involved.

PUTTING DECISIONS INTO PRACTICE: A MORE COHERENT AND SYSTEMATIC APPROACH

Improving the existing processes

35. No new process is needed. The existing Broad Economy Policy Guidelines and the Luxembourg, Cardiff and Cologne processes offer the necessary instruments, provided they are simplified and better coordinated, in particular through other Council formations contributing to the preparation by the ECOFIN Council of the Broad Economic Policy Guidelines. Moreover, the Broad Economic Policy Guidelines should focus increasingly on the medium- and long-term implications of structural policies and on reforms aimed at promoting economic growth potential, employment and social cohesion, as well as on the transition towards a knowledge-based economy. The Cardiff and Luxembourg processes will make it possible to deal with their respective subject matters in greater detail.

36. These improvements will be underpinned by the European Council taking on a pre-eminent guiding and coordinating role to ensure overall coherence and the effective monitoring of progress towards the new strategic goal. The European Council will accordingly hold a meeting every Spring devoted to economic and social questions. Work should consequently be organised both upstream and downstream from that meeting. The European Council invites the Commission to draw up an annual synthesis report on progress on the basis of structural indicators to be agreed relating to employment, innovation, economic reform and social cohesion.

Implementing a new open method of coordination

37. Implementation of the strategic goal will be facilitated by applying a new open method of coordination as the means of spreading best

practice and achieving greater convergence towards the main EU goals. This method, which is designed to help Member States to progressively develop their own policies, involves:

- fixing guidelines for the Union combined with specific timetables for achieving the goals which they set in the short, medium and long terms;
- establishing, where appropriate, quantitative and qualitative indicators and benchmarks against the best in the world and tailored to the needs of different Member States and sectors as a means of comparing best practice;
- translating these European guidelines into national and regional policies by setting specific targets and adopting measures, taking into account national and regional differences;
- periodic monitoring, evaluation and peer review organised as mutual learning processes.

38. A fully decentralised approach will be applied in line with the principle of subsidiarity in which the Union, the Member States, the regional and local levels, as well as the social partners and civil society, will be actively involved, using variable forms of partnership. A method of benchmarking best practices on managing change will be devised by the European Commission networking with different providers and users, namely the social partners, companies and NGOs.

39. The European Council makes a special appeal to companies' corporate sense of social responsibility regarding best practices on lifelong learning, work organisation, equal opportunities, social inclusion and sustainable development.

40. A High Level Forum, bringing together the Union institutions and bodies and the social partners, will be held in June to take stock of the Luxembourg, Cardiff and Cologne processes and of the contributions of the various actors to enhancing the content of the European Employment Pact.

Mobilising the necessary means

41. Achieving the new strategic goal will rely primarily on the private sector, as well as on public–private partnerships. It will depend on mobilising the resources available on the markets, as well as on efforts by Member States. The Union's role is to act as a catalyst in this process, by establishing an effective framework for mobilising all available resources for the transition to the knowledge-based

economy and by adding its own contribution to this effort under existing Community policies while respecting Agenda 2000. Furthermore, the European Council welcomes the contribution that the EIB stands ready to make in the areas of human capital formation, SMEs and entrepreneurship, R&D, networks in the information technology and telecom sectors, and innovation. With the 'Innovation 2000 Initiative', the EIB should go ahead with its plans to make another billion euro available for venture capital operations for SMEs and its dedicated lending programme of 12 to 15 billion euro over the next 3 years for the priority areas.

Documents submitted to the Lisbon European Council[1]

Presidency note on Employment, Economic Reforms and Social Cohesion
Towards a Europe based on Innovation and Knowledge
(5256/00 + ADD1 COR 1 (en))

Commission report
*e*Europe – An information society for all
(6978/00)

Commission contribution
– An Agenda of economic and social renewal for Europe
(6602/00)

Commission communication on Community policies in support of
employment
(6714/00)

Commission communication. Building an inclusive Europe
(6715/00)

Commission communication. Social trends: prospects and challenges
(6716/00)

Commission communication. Strategies for jobs in the information society
(6193/00)

Commission report on economic reform. Report on the functioning of
product and capital markets
(5795/00)

Contribution of the Council (ECOFIN)
(6631/1/00 REV 1)

Contribution of the Council (Labour and Social Affairs)
(6966/00)

---

[1] The preparatory documents concerning employment, economic reform and social cohesion
can be found on the Presidency's internet site at: http://www.portugal.ue-2000.pt/

Contribution of the Council (Internal Market) Cardiff Economic Reform
Process: Internal Market Aspects
(7130/00)

Opinion of the Employment and Labour Market Committee
(6557/00)

Presidency report 'Strengthening the Common European Security and
Defence Policy'
(6933/00)

Report on the Western Balkans presented to the European Council by the
Secretary General/High Representative together with the Commission
(SN 2032/2/00 REV 2)

Draft report from the European Council to the European Parliament on
the progress achieved by the European Union in 1999
(6648/00 + COR 1 (gr))

Presidency Conclusions

Santa Maria da Feira European Council
19 and 20 June 2000

(Paragraphs concerning Lisbon Special European Council Follow-up)

---

1. The European Council met in Santa Maria da Feira on 19 and 20 June. At the start of proceedings, the European Council and the President of the European Parliament, Mrs Nicole Fontaine, exchanged views on the main items under discussion.

II. EMPLOYMENT, ECONOMIC REFORMS AND SOCIAL COHESION – FOLLOW-UP TO THE LISBON EUROPEAN COUNCIL

19. The Lisbon strategy, which is now fully underway, underpins all Community action for jobs, innovation, economic reform and social cohesion. Substantial results are already being delivered on all areas covered by the strategy.

20. The High Level Forum bringing together the social partners, the Union's institutions, the European Central Bank and the European Investment Bank met in Brussels on 15 June. It confirmed that there was a high degree of consensus on the Lisbon strategy, identified the possible contributions of the various actors, each within its own sphere of action, and demonstrated the importance of broad political debate, social concertation and social dialogue. More particularly, the European Council welcomes the Joint Declaration presented by the social partners, which sets out constructive positions on temporary work, telework, lifelong learning and provisions for joint monitoring of industrial change.

21. Momentum in implementing the strategy must be sustained by fixing the next priority steps outlined below.

A.   Preparing the transition to a competitive, dynamic and knowledge-based economy

eEurope Action Plan

22.   The European Council endorses the comprehensive *e*Europe 2002 Action Plan and requests the institutions, the Member States and all other actors to ensure its full and timely implementation by 2002 and to prepare longer-term perspectives for a knowledge-based economy encouraging info-inclusion and closing the numeracy gap. As a short-term priority, the necessary steps should be taken to bring down the cost of accessing the Internet through the unbundling of the local loop. A report should be presented by the Commission to the European Council in Nice, and on a regular basis thereafter, on progress in achieving the Actions Plan's objectives. The European Council recalls the strategic importance of the Galileo project and of taking a decision on this matter by the end of 2000.

The creation of a European Research Area

23.   The European Council welcomes the resolution by the Council (Research) and the commitment rapidly to develop a 'European Research Area' in particular by drawing up criteria for benchmarking research policies; defining stages and deadlines for mapping scientific and technological excellence in Europe by 2001; undertaking to network national and European research programmes; and taking initiatives to interconnect at very high speed national electronic networks for research. Following the conclusions of the last US-EU Summit, the Commission is invited to pursue actively the dialogue with the US authorities to establish a broadband, permanent and equitable transatlantic link between European and US research and education centres.

The European Charter for Small Enterprises and the new framework for enterprise policy

24.   The European Council welcomes the recently adopted European Charter for Small Enterprises *(see Annex III)*, and underlines the importance of small firms and entrepreneurs for growth, competitiveness and employment in the Union. It requests its full implementation as part of the comprehensive framework for enterprise policy under preparation. This comprehensive framework is advancing on the basis of the Commission's proposed Work Programme for Enterprise Policy 2000–2005, the proposals for benchmarking enterprise policy as well as the proposed Multiannual Programme for Enterprise and

Entrepreneurship 2001–5. First results of these exercises must be achieved by the end of 2000.

Completing the Internal Market

25. The European Council endorses the general orientation contained in the Commission's Communication on the Review of the Internal Market Strategy as a useful basis for planning work. A coherent framework is needed to develop the internal market by aligning future reviews of the Commission's Strategy with the Cardiff economic reform process so that the Spring European Councils can fully assess progress.

26. Significant steps towards completing the internal market have already been taken. Thus

- the directive on electronic commerce has been adopted, and agreement reached on harmonising certain aspects of copyright and related rights;
- under the Financial Services Action Plan, political agreement has been achieved on common positions on take-over bids, the winding-up and reorganisation of credit institutions and insurance undertakings; furthermore, the e-money directive has just been adopted;
- the Commission has presented proposals for new rules for public procurement and for the next stage of postal liberalisation; it will shortly present further proposals for reform of the telecoms sector and for a Community patent and the utility model.

27. Work now needs to be carried forward on other aspects. The Commission is accordingly:

- invited to present a report by March 2001 on the evolution of energy markets according to the Lisbon strategy;
- called on to continue the work of the High Level Group on a single European sky in order to present a final report in the first half of 2001 with a view to bringing forward appropriate proposals.

28. The European emphasised that the concerns and significance of public services of general interest must be taken into account in a dynamic single market. In this connection, it reiterated its request to the Commission to update its 1996 communication on public services

of general interest. It expects that the updated communication will be submitted by its next meeting in Biarritz at the latest.

29. The Council and the European Parliament are invited to speed up work on the money laundering directive and the UCITS directive. The Commission is urged to present proposals for a single licence for issuers (prospectus directive), a new accounting strategy to enhance the comparability of listed companies' financial statements and measures to promote consumer confidence in the field of financial services, including distance marketing and e-commerce.

30. The European Council welcomes the rapid implementation of the 'Innovation 2000 Initiative' by the European Investment Bank and its contribution to developing a knowledge-based economy and social cohesion. It calls on the Bank to pursue its efforts in cooperation with national and regional authorities, the financial community and the Commission.

31. The European Council stresses the role of public administrations, administrative action and better regulation in enhancing the competitiveness of the Union and of the Member States, thus contributing to economic growth and employment opportunities. The European Council encourages Member States to review the quality and performance of public administration in view of the definition of a European system of benchmarking and best practices.

B. Modernising the European social model by investing in people and building an active welfare state

Education and training for living and working in the knowledge society

32. The European Council welcomes the Commission's communication on e-learning and endorses the Council guidelines on future challenges and objectives of education systems in the learning society. These provide a framework for preparing the broader report on education to be presented in the European Council in Spring 2001, as well as a method for improving the contribution of education policies to the Luxembourg process.

33. Lifelong learning is an essential policy for the development of citizenship, social cohesion and employment. The Member States, the Council and the Commission are invited, within their areas of competence, to identify coherent strategies and practical measures with a view to fostering lifelong learning for all, to promote the involvement of social partners, to harness the full potential of public and private financing, and to make higher education more accessible to more people as part of a lifelong learning strategy.

Developing the active employment policy

34. The Union is enjoying substantially improved employment prospects. Efforts to further strengthen the European Employment Strategy through the mid-term review of the Luxembourg process should underpin the revision of the Employment Guidelines for 2001. In this context, the social partners are invited to play a more prominent role in defining, implementing and evaluating the employment guidelines which depend on them, focusing particularly on modernising work organisation, lifelong learning and increasing the employment rate, particularly for women.

Modernising social protection, promoting social inclusion

35. A number of priorities have already been identified in this area:

- as regards the future evolution of social protection, particular attention should be given to the sustainability of pension schemes through defining two action lines aimed at improved forecasting of future trends and at obtaining in-depth knowledge of recent, actual or expected national pension reform strategies;
- on promoting social inclusion, a framework with appropriate objectives should be defined to evaluate the impact of social policies applied in Member States and indicators should be defined as common references in the fight against social exclusion and the eradication of poverty.

36. Development and systematic monitoring of work on these matters at Community level will be improved by the recent setting up the Social Protection Committee, regular debate on those issues and by encouraging cooperation between Member States through an open method of coordination combining national action plans with a Community programme to combat social exclusion. On this latter point, the Council is invited to adopt rapidly the Commission's recent proposal for this programme. Appropriate association of the social partners with the ongoing work should also be developed. The conclusions of the Lisbon European Council made a special appeal to companies' corporate sense of social responsibility. The European Council notes with satisfaction the ongoing follow up to this and welcomes the initiation of the process to establish a network for a European dialogue on encouraging companies' corporate sense of social responsibility. The European Council notes with satisfaction the recent political agreement reached in the Council on a directive establishing a legal

framework for combating discrimination on the grounds of racial or ethnic origin.

The European Social Agenda

37. The draft European Social Agenda proposed by the Commission will provide a multi-annual framework for action in social matters. The Council is invited to examine the Agenda as a matter of urgency to enable it to be endorsed by the Nice European Council.

Improving working methods

38. Implementation, monitoring of progress and follow-up of the Lisbon strategy will take place within the existing institutional framework and will be consolidated by:

- improving coordination between the various Council forma-tions and ensuring close cooperation between the Council Presidency and the Commission, under the overall guidance of the European Council, in line with the recommendations approved at the Helsinki European Council;
- developing and improving from a methodological point of view the open method of coordination where appropriate within the framework of the Council as one of the possible instruments in policy fields such as information society, research, innovation, enterprise policy, economic reforms, education, employment and social inclusion;
- the Commission presenting a report by the end of September on the proposed approach for indicators and benchmarks, both in specific policies and to be used in the synthesis report to the Spring European Council, to ensure the necessary coherence and standard presentation.

D. The Spring European Council

39. The European Council looks forward to holding its first regular Spring session on economic and social strategy and policies in Stockholm early in 2001 based on the annual synthesis report to be presented by the Commission and taking into account the contribu-tions of the various Council formations. The social partners should also have the opportunity to address these issues before the European Council.

II. ECONOMIC, FINANCIAL AND MONETARY AFFAIRS

Broad Economic Policy Guidelines

40. The European Council welcomes the 2000 Broad Economic Policy Guidelines for the Community and the Member States. As reiterated at the Lisbon European Council, they are designed to maintain growth and stability-oriented macroeconomic policies, increase the growth potential of the Union, improve the quality and sustainability of public finances and move forward with far-reaching and comprehensive reforms of product, capital and labour markets. They build on the Lisbon strategy, thereby contributing to meeting, over time, the challenges of restoring full employment, promoting the transition to a knowledge-based economy, preparing for the consequences of ageing populations and improving social cohesion. They also improve the synergies between Cologne, Cardiff and Luxembourg processes.

41. The European Council invites the ECOFIN Council to implement its conclusions on practical steps forward to enhance the coordinating role of the Broad Economic Policy Guidelines in close cooperation with the other relevant Council formations and by addressing the link between structural and macroeconomic policies. Macroeconomic dialogue should also be improved.

B. Tax package

42. The European Council endorses the report on the tax package by the ECOFIN Council *(see Annex IV)*, the statements included for the Council minutes and the agreement on its principles and guidelines. It endorses the timetable set out, which foresees a step-by-step development towards realisation of the exchange of information as the basis for the taxation of savings income of non-residents. The European Council requests the ECOFIN Council to pursue with determination work on all parts of the tax package so as to achieve full agreement on the adoption of the directives and the implementation of the tax package as a whole as soon as possible and no later than by the end of 2002.

Bibliography

DOCUMENTS CONNECTED WITH LISBON EUROPEAN COUNCIL

Information Society

European Commission, 'eEurope an Information Society for all – Progress Report', Ref. COM (2000) 130 final, 08.05.00.

European Commission, 'eEurope 2002 An Information Society for all, Action Plan prepared for Feira European Council', 14.06.00.

European Commission, 'Amended Proposal for a Regulation of the European Parliament and of the Council on unbundled access to the local loop', Ref. COM (2000) 761, 22.11.00.

Council of the European Union, 'Bilan des actions menées par la Présidence pour la mise en oeuvre du Plan d'action eEurope', Ref. Council of the European Union 13515/00, 24.11.00.

Council of the European Union, 'Bilan des actions menées par la Présidence pour la mise en oeuvre du Plan d'action eEurope – Contributions des Etats membres', Ref. Council of the European Union 13515/00 ADD 1, 27.11.00.

Research

European Commission, 'Communication: Towards a European Research Area', Ref. COM (2000) 06, 18.01.00.

Presidency of the European Union, 'Revised Presidency Proposal for a draft Council Resolution on establishing a European area of research and innovation', 15.05.00.

European Commission, 'Communication: Making a reality of the European Research Area: Guidelines for EU research activities (2002–2006)', Ref. COM (2000) 612 final, 4.10.00.

European Commission, 'Communication: Comments of the Commission on the conclusions of the RTD Framework Programmes 5-Year Assessment', Ref. COM (2000) 659, 19.10.00.

European Commission, 'Elaboration of an open method of coordination for the benchmarking of national research policies – objectives, methodology and indicators', Ref. SEC (2000) 1842, 31.10.00.

Council of the European Union, 'Projet de résolution du Conseil sur la réalisation de l'Espace européen de la Recherche et de l'Innovation: orientations pour les actions de l'Union en matière de recherche (2002–2006)', Ref. 12881/00, 10.11.00.

Innovation

European Commission, 'Communication: Innovation in a knowledge-driven economy', Ref. COM (2000) 567 final, 20.09.2000.

European Commission, 'Working Paper: Trends in European innovation policy and the climate for innovation in the Union', Ref. SEC (2000) 1564.

Enterprise Policy

European Commission, 'Towards Enterprise Europe – Work Programme for enterprise policy 2000–2005', 18.04.2000.

European Commission, 'Communication from the Commission Challenges for Enterprise policy in the knowledge-based economy, Proposal for a Council Decision on a multiannual Programme for Enterprise and Entrepreneurship (2001–2005)', Ref. COM (2000)256 final, 26.04.2000.

European Commission, 'Benchmarking Enterprise Policy', 05.05.2000.

European Commission, 'European Charter for Small Enterprises', Ref. Council of the European Union 9331/00, 09.06.00.

Technical Note supporting the Charter for Small Enterprises

European Commission, 'Communication: Review of specific Community financial instruments for SMEs', Ref. COM (2000) 653, 20.10.00.

European Commission, 'Benchmarking Enterprise Policy – First results from the Scoreboard', Ref. SEC (2000) 1842, 27.10.00.

European Commission, 'European competitiveness report 2000', Ref. SEC (2000) 1823, 30.10.00.

Education

European Commission, 'Communication from the Commission eLearning – Designing tomorrow's education', Ref. COM (2000) 318 final, 24.05.00.

Presidency of the European Union, 'Future challenges and objectives of education systems in the learning society: follow-up to the Lisbon European Council (Background paper for debate)', Ref. Council of the European Union 8880/00, 26.05.00.

European Commission, 'Amended proposal for a Recommendation of the European Parliament and of the Council on mobility within the Community for students, persons undergoing training, young volunteers, teachers and trainers', Ref. COM (2000) 723 final, 09.11.00.

European Commission, 'Memorandum on education and lifelong learning'.

European Commission, 'Report on the future objectives of education systems'.

Employment

European Commission, 'Communication: Strategies for jobs in the Information Society', Ref. COM (2000) 48 final, 04.02.00.

European Commission, 'Communication: European Policies in support of employment', 2000.

Employment and Labour Market Committee, 'Opinion with a view to the Lisbon Special European Council', Ref. Council of the European Union 6557/00, 06.03.00.

Council of the European Union, 'Contribution from the Council (Labour and Social Affairs) for Lisbon Summit', Ref. Council of the European Union 6966/00, 17.03.00.

Council of the European Union (Employment and Social Policy), 'Conclusions on the follow-up of Lisbon European Council', Ref. Council of the European Union 9353/00, 09.06.00.

Employment package

European Commission, 'Joint Employment report 2000', Ref. COM (2000) 551, September 2000.

European Commission, 'Recommendation for a Council Recommendation on the implementation of Member States' employment policies', Ref. COM (2000) 549, September 2000.

European Commission, 'Proposal for a Council Decision on Guidelines for Member States' employment policies for the year 2001', September 2000.

European Commission, 'EURES activity report 1998–99 "Towards an integrated European labour market: the contribution of EURES"', Ref. COM (2000) 607 final, 02.10.2000.

European Commission, 'Amended Proposal for a Council Directive establishing a general framework for equal treatment in employment and occupation', Ref. COM (2000) 652 final, 12.10.00.

Employment Committee and Economic Policy Committee, 'Joint Opinion of the Employment Committee and the Economic Policy Committee on the Commission proposal for a Council Decision on guidelines for member States' employment policies for the year 2001', Ref. Council of the European Union 12908/00, 6.11.00.

Economic and financial aspects

European Commission, 'Report on the functioning of European product and capital markets', Ref. COM (2000) 26 final, 26.01.00.

Council of the European Union, 'Contribution from the Council (ECOFIN) for the Special European Council on Employment, Economic Reform and Social Cohesion', Ref. Council of the European Union 6631/1/00, 13.03.00.

European Investment Bank, Ref. Council of the European Union 6442/00, 15.03.00

Council of the European Union, 'Conclusions of the Council (Internal Market)', Ref. Council of the European Union 7130/00, 20.03.00.

Council of the European Union, 'Commission Recommendation on the 2000 Broad Guidelines on the Economic Policies of the Member States and the Community', Ref. COM (2000) 214 final, 11.04.00.

European Commission, Financial Services Action Plan – 'Second Progress Report', Ref. COM (2000) 336 final, 30.05.00.

European Investment Bank, 'From Lisbon to Santa Maria da Feira – Progress in implementing the EIB's "Innovation 2000 Initiative"'

Council of the European Union (Ecofin), 'Tax Package Report', Ref. Council of the European Union 9034/00, 20.06.00.

European Commission, 'Communication: Services of general interest in Europe', Ref. COM (2000) 580 final, 20.09.00.

European Commission, 'Communication: Progress Report on the Risk Capital Action Plan', Ref. COM (2000) 658 final, 18.10.00.

European Commission, 'Financial services priorities and progress – Third Report', Ref. COM (2000) 692 final, 08.11.00.

Social protection and social inclusion

European Commission, 'Communication: Social Trends Prospects and Challenges', Ref. COM (2000) 82 final, 01.03.00.

European Commission, 'Communication: Building an inclusive Europe', Ref. COM (2000) 79 final, 01.03.00.

High Level Party on Social Protection, 'Report of the High Level Party on Social Protection relating to the effort of the co-operation for the modernisation and the enhancement of social protection', Ref. 8634/00, 18.05.00.

European Commission, 'Proposal for a European Parliament and Council

Decision establishing a Community Action Plan to encourage co-operation between the Member States to fight social exclusion', Ref. COM (2000) 368 final, 16.06.00.

European Commission, 'Amended Proposal for a Council Decision establishing a Community Action Programme to combat discrimination 2001–2006', Ref. COM (2000) 649 final, 10.10.00.

COREPER, 'Fight against poverty and social exclusion – Definition of objectives for Nice European Council – Political agreement', Ref. Council of the European Union 12189/00, 10.10.00.

European Commission, 'Communication: The future evolution of Social Protection from a Long-Term Point of view: Safe and Sustainable Pensions', Ref. COM (2000) 622, 24.10.00.

Council of the European Union, 'EPC progress report on the impact of ageing populations on public pensions systems', Ref. Council of the European Union 12791/00, 30.10.00.

High-Level Working Party on Social Protection, 'Progress Report on the Commission Communication on the future evolution of Social Protection from a Long-Term Point of view: Safe and Sustainable Pensions', Ref. Council of the European Union 12949/00, 06.11.00.

European Commission, 'Amended Proposal for a Decision of the European parliament and of the Council establishing a programme of Community action to encourage cooperation between Member States to combat social exclusion', Ref. COM (2000) 796, 24.11.00.

European Commission, 'Amended Proposal for a Council Decision on the Programme relating to the Community framework strategy on gender equality', Ref. COM (2000) 793, 24.11.00.

Social dialogue

ETUC, 'ETUC Memorandum', Ref. Council of the European Union 6685, 03.03.00

UNICE, Ref. Council of the European Union 6686, 09.03.00.

Standing Committee on Employment, Presidency Conclusions, Ref. ESC 501/1/00 REV 1, 13.03.00.

Eurocommerce, Ref. Council of the European Union 6957/00, 15.03.00.

Presidency of the European Union, 'Presidency's Summing-up of the debate during the High Level Forum', 15.06.00.

ETUC, UNICE, CEEP, 'Joint Statement of Social Partners to the High Level Forum', 15.06.00.

Transverse matters

Presidency of the European Union, 'Document from the Presidency', Ref. Council of the European Union 5256/00, 12.01.00.

Presidency of the European Union, 'International Hearing for the Portuguese Presidency of the European Union', Lisbon, 3–4 December 1999.

Presidency of the European Union, 'Reports prepared for the Portuguese Presidency of the European Union'.

European Commission, 'Contribution for Lisbon Special European Council: An Agenda of Economic and Social Renewal for Europe', Ref. CG1 (2000) 4, 22.02.00

Economic Policy Committee, 'Annual Report on Structural Reforms 2000', Ref. Council of the European Union 6655/1/00 REV 1, 15.03.00.

European Parliament, 'Resolution of the European Parliament on the Lisbon Special European Council', Ref. B5-0236,0239 and 0240/2000.

Presidency Conclusions (Lisbon European Council), Ref. Council of the European Union SN 100/00.

Presidency of the European Union, 'Follow-up of Lisbon European Council Conclusions', Ref. 7953/00, 19.04.00.

European Parliament, 'Motion for a resolution – European Parliament Resolution on the preparation of the European Council meeting in Feira on 19 and 20 June 2000', Ref. B5-0529/2000.

Presidency of the European Union, 'Note on the ongoing experience of the open method of co-ordination', 9088/00, 14.06.00.

Presidency Conclusions (Santa Maria da Feira European Council), Ref. Council of the European Union SN 200/00.

European Commission, 'Communication: Social Policy Agenda', Ref. COM (2000) 379 final, 28.06.2000 (Presented in the end of the Portuguese Presidency).

High-Level Working Party on Social Protection, 'Opinion of the High-Level Working Party on Social Protection – Social Policy Agenda', Ref. Council of the European Union 12772, 27.10.00.

Employment Committee, 'Opinion of the Employment Committee on the Social Policy Agenda', Ref. Council of the European Union 12988/00, 6.11.00.

Council of the European union, 'European Social Agenda – Contribution by the Council (Employment and Social Policy) to the Nice European Council with a view to the adoption of a European Social Agenda', Ref. Council of the European Union 13880/1/00, 28.11.00.

European Commission, 'Communication: Structural indicators', Ref. COM (2000) 594, 27.09.2000.

Economic Policy Committee, 'Report by the Economic Policy Committee to ECOFIN on "Structural Indicators: an instrument for Better Structural Policies"', Ref. EPC/ECFIN/608/00-fin, 26.10.00.

Council of the European Union (ECOFIN), 'Report by the Council (ECOFIN) to the European Council in Nice on "Structural Indicators: an instrument for Better Structural Policies"', Ref. Council of the European Union 13217/00, 27.11.00

Presidency Conclusions (Nice European Council), Ref. Council of the European Union SN 400/00.

Index